顶级设计空间
TOP DESIGN SPACE

《顶级设计空间》编委会 编

潮流商铺
FASHION STORES

专卖店 & 商业展示
STORES & COMMERCE SHOWS

（第二版）

中国林业出版社

潮流商铺：汉英对照 /《顶级设计空间》编委会编
. -- 2版. -- 北京：中国林业出版社，2013.4
（顶级设计空间）

ISBN 978-7-5038-7273-0

Ⅰ. ①潮… Ⅱ. ①顶… Ⅲ. ①商店－室内装饰设计－
图集 Ⅳ. ①TU247.2-64

中国版本图书馆CIP数据核字(2013)第274861号

《顶级设计空间》编委会编
主编：张青萍
编委：孔新民、贾陈陈、许科、李钢、吴韵、竺智、曾丽娴

责任编辑：纪亮
英文翻译：董君、梅建平、牛晓霆、万毅、赵强

出版：中国林业出版社
　　　（100009 北京西城区德内大街刘海胡同7号）
网址：http://lycb.forestry.gov.cn/
E-mail：cfphz@public.bta.net.cn
电话：(010) 83143581
发行：新华书店
印刷：北京卡乐富印刷有限公司
版次：2016年4月第2版
印次：2016年4月第1次
开本：230mm×300mm
印张：32
字数：360千字
定价：199.00元

设计进行时

中国改革开放30年，室内设计行业行进到今天，也已经有了无计其数的变化、发展和积累，30年的思考、30年的实践、30年的进步，也造就了这30年的成绩。

我们好似在进行着一场接力赛，祖先把中华民族灿烂的文化一代代地传承到21世纪，我们有责任将这份优秀的遗产倍加珍藏以传给后代。我们所面临的挑战是拿什么当代的室内文化馈赠后人？但其实这30年中由于电子技术的普及和信息的迅速传播和交换，设计已出现国际化、同一化的倾向，与此同时引起了传统性、地域性和个性差异的不断丧失，又有由于社会追求物质与功能价值的同时造成对精神和文化价值的忽视，我们已找不到回头的路。但不管历史结论会如何，我们这代人是努力的、勤奋的，是不断地用自己的智慧为中国室内设计行业进步奉献着的。

总的看来，21世纪的室内设计发展有以下倾向和趋势：

倡导绿色设计

人类起源于自然，其间虽曾摆脱过自然，但最终还是要以全新的面目去回归于自然。如此轮回恰恰历史地、辩证地道出了人与自然关系的变化。如今的人们，特别是生活在大城市里的人离大自然已越来越远了，于是人们特别希望在室内再现一些大自然的情景，以求得哪怕是暂时地、局部地享受。作为设计师一方面尽可能地创造出生态环境，让人们最大限度地接近自然，另一方面须有环保意识，努力去提高设计中的健康因素，以满足人们在生理和心理上的需要。

室内设计中的健康设计充分利用自然或仿自然的因素，为人们提供生活舒适的空间。室内的色彩、照明及功能空间的弹性分割，都应该在满足其基本功能的基础上，尽可能充分利用自然能源。尤其是提倡对绿色装饰材料和绿色照明材料的运用，同时注重社会心理学的研究。绿色设计本着以下几个原则：

①设计上使用最少的材料，少浪费应节约的资源、能源，力戒奢靡；②尽量多采用污染少，环保性能强，安全可靠的材料；③符合人体工程学的要求，讲求空间上比例与尺度，避免使人感到压抑或繁琐；④设计应以人为本，满足人的本质需要。

运用高科技手段

科技进步影响着人类生活的方方面面，现代科技的发展为人类的衣食住行提供了很多方便，它可以使居住环境更加符合人们的意愿。因此，在设计中提高科技含量，创造高效率、高功能、高质量室内生活环境的要求已愈加鲜明。

在本世纪，楼宇的智能化将逐步实现。建筑智能化就是将结构、设备、服务运营及相互关系进行全面综合配置，从而达到最佳的组合程度，使建筑具有高效率、高功能和舒适性。单从自动化来看，就是实现建筑设备自动化、办公自动化和通讯自动化。建筑智能化并非仅仅针对大楼宇，还会迅速走向寻常百姓家。住宅中自动防盗报警，自动调温、调湿、自动除尘、调节灯光亮度、自动控制炊事用具等，如今也已成为事实。

注重设计的系统性

在这个日新月异、急剧变化的时代，系统地看待问题和解决问题是当代人的特质。站在现代与历史之间的人们既希望从传统中找回精神的家园，以弥补快速发展带来的心理失落与不安；同时又满怀着激情和野心试图运用当代技术和审美重新诠释历史，使之适应现代生活。系统的眼光可使我们将面临的室内设计浅层次问题渗透到更深的层次方面加以科学、综合地解决。这样的设计系统性包括环境学、生态学、经济学、系统论、方法论、控制论、统筹学、管理学及有关室内设计方面的政策法规、标准规范等方面的内容。

室内设计系统是指应用系统的观点和方法，将室内设计的内容、要素，相关的领域和环节，以及室内设计的程序予以统筹而形成的一个框架体系。从与其相关部分的关系和进行的程序来分析，可理解为有横向设计系统和纵向设计系统两个方面。横向系统设计表现为在设计过程中所涉及到的如生理学、心理学、行为科学、人体工程学、材料学、声学、光学、经济学等诸多因素；纵向系统设计表现为对设计实现过程中所有历程的考虑。概括而言，横向系统设计强调相关与联系；纵向系统设计强调过程与变化。

无论将来的室内设计如何发展，它都必然在一个更广泛、更全面的系统里科学地伸展，设计师也必将持有这种科学的态度和掌握这类理性的方法而从事设计。

本系列图书刊登的是近年来一些顶级的设计作品，它们或多或少的反映的是当下室内设计师的思考和当前技术背景下的实践。30年告了一个段落，下一个30年又将开始，我们已走上新的征程，设计永远是进行时。愿这本套书的出版能得到业界的认可和赞扬！

南京林业大学 风景园林学院副院长、教授
id+c《室内设计与装修》杂志主编　张青萍
2010.3.1草于南京

专卖店

1. Club Designer旗舰店 8
2. Ayres商店 16
3. Asobio上海旗舰店 22
4. 香港镇金店 28
5. 和成卫浴广州旗舰店 34
6. Galeries Lafayette店铺 42
7. Kiki 2专卖店 48
8. Geometry男装专卖店 54
9. 韩国某服饰店 60
10. Jean-Patou香水店 70
11. Bastard旗舰店 74
12. Novo 北京服装店 82
13. Novo上海服装店 86
14. Custo巴塞罗那店 90
15. Villa Moda巴林店 94
16. 杭州浪漫一身 100
17. 捷安特女性旗舰店 106
18. Kymyka鞋包店 110
19. Pamper Heiress旗舰店 114
20. Baci & Abbracci 旗舰店 120

23. 上海蒲蒲兰绘本馆 134
24. Marni品牌旗舰店 140
25. One2free旗舰店 146
26. Vilasofa旗舰店设计 152
27. Linden Apotheke药店 158
28. s.Oliver形象概念店铺 164
29. Asobio店铺设计 172
30. SSSP 唱片连锁店 178
31. 芭比上海旗舰店 182

商业展示

32. 上海空间美学馆 194
33. Isoleé时尚生活馆 202
34. 开元食品展示中心 210
35. Vitra家具零售商店设计 216
36. 广州本田企业形象展示厅 224
37. 北京木皇家具 230
38. Lex百货公司 238
39. 雪铁龙C_42巴黎旗舰店 244
40. 弘第厨具台北展馆 252

Stores

1. CLUB DESIGNER 8
2. INTERIOR DESIGN OF AYRES STORE 16
3. ASOBIO FLAGSHIP STORE 22
4. DESIGN OF JUSTGOLD 28
5. HCG FLAGSHIP IN GUANGZHOU 34
6. GALERIES LAFAYETTE SHOP 42
7. INTERIOR DESIGN OF KIKI 2 SHOP 48

23. POPLAR KID'S REPUBLIC 134
24. MARNI FLAGSHIP 140
25. ONE2FREE FLAGSHIP SHOP 146
26. VILASOFA FLAGSHIP STORE 152
27. LINDEN APOTHEKE 158
28. LINE ELEMENTSLINDEN IN S.OLIVER 164
29. ASOBIO SHOP IN SHANGHAI 172
30. SIRIUS SMART SOUNDS PRAGUE 178

STO

CLUB DESIGNER
Club Designer 旗舰店

01

【坐落地点】中国台湾台北市大安路一段133号1楼；【面积】238 m²
【设计】李玮珉；【设计公司】LWMA李玮珉建筑师事务所
【主要建材】墨镜、镜面不锈钢、盘多魔地板、皮革、灯泡、透光深紫马赛克、玻璃
【摄影】李国民

设计师巧妙地运用了灯泡作为一楼外观空间元素的一部分，特别量身订制珍珠玻璃外壳灯泡镶嵌在黑墙上，带有复古感亦能透过设计安排之后的灯光明灭，呈现该店的时尚讯息。外立面并未被灯泡墙完全遮挡，半幅的玻璃墙面让室内空间的LED主墙面恰好与外面的灯泡墙在同一个平面上衔接，室内与室外融合在一起。设计师采用反向思考的方式呈现设计概念，将所有的客户遮蔽，在街上几乎看不到店内的活动，只看得到品牌形象。

室内整体以墨镜为主，一层的天花全部为镜面不锈钢，拉伸了原本不高的空间。天花线条一直延续至楼梯的主墙面，并转换为柜台后方凹凸的机能柜，成为利落的空间元素。一层的LED主墙一直延续到地下一层，LED的排列方式随时转化，优雅的菱格纹可以变幻成其他各种店主需要的花纹。

地下一层在设计手法上较活泼。透光深紫马赛克与柱的元素相结合，通过墨镜及镜面不锈钢的反射，使其更有延伸的张力。镶嵌在黑镜后的光纤条变换成各种颜色，跳动在空间中，无形中分隔了不同的品牌。空间中深凹的红色凹洞暗示更衣室的入口，楼梯左侧银紫色拉帘，弹性区隔的VIP区及展示区，搭配全室略带紫灰的地板，让Club Designer散发出无与伦比的时尚吸引力。

Designers skillfully used the lamp as a part of the first floor of the appearance of space elements, in particular, tailored pearl inlaid in black glass bulb shell wall, with a retro feel can also arrange, through the design after the lights flicker, showing the store fashion message. Light bulb wall facade has not been completely blocked, demi glass wall so that interior space of the main wall LED light bulb with the outside wall exactly in the same plane convergence, indoor and outdoor together. Designers to reverse ways of thinking presented the design concept, all customer shelter on the street hardly see the store's activity, only its brand image.

Sunglasses indoors as a whole to the main floor ceiling entirely of stainless steel mirror, stretching the original space is not high. Smallpox line has been extended to the main staircase, walls, and converted to the function of cabinets behind the counter to bump into the space element of neat. Layer of the LED of the main wall has been extended to the basement, LED of the arrangement at any time into an elegant diamond-shaped patterns may be changing into the other shop owners need to tread.

Basement design methods in a more lively. Mosaic combines elements of the column, through the dark glasses and the reflection mirror stainless steel, making it more an extension of the tension. Embedded in the black section of fiber mirror transform into a variety of colors, beating in the space, virtually separated by a different brand. Deep in space the red pits suggest locker room entrances, staircases on the left Silver purple curtain pull, flexible segment of the VIP area and display area, with the whole room slightly purple on the floor, so that exudes unparalleled Club Designer fashion to attract force.

↑ 一层以沉稳作为空间的基调 / Layer in order to calm the tone as space.

↑ 镜面不锈钢天花将空间延伸 / Mirror stainless steel ceiling space extension.
← 订制的珍珠玻璃外壳在夜晚如繁星闪耀 / Custom glass pearl shell, such as the stars shine at night.

↑ 深凹的红色凹洞+暗示更衣室的入口 / Deep red pits suggest locker room entrance.
← 地下一层的不同角度 / Basement of the different angles.
↓ 地下一层平面图 / Basement floor plan; ↓ 一层平面图 / 1st floor plan.

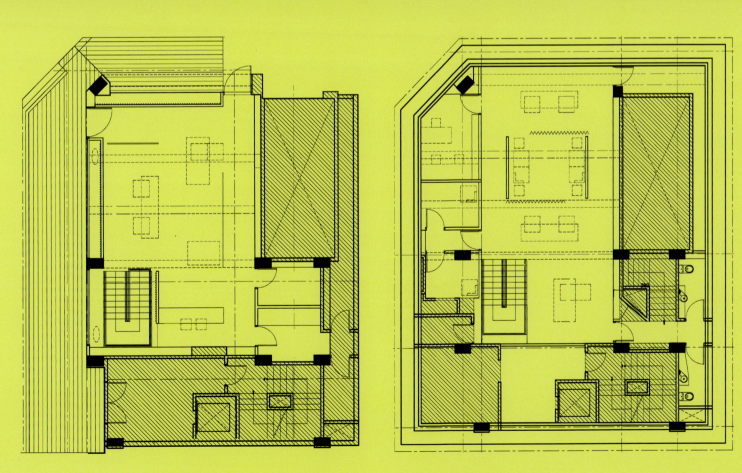

INTERIOR DESIGN OF AYRES STORE

Ayres 商店

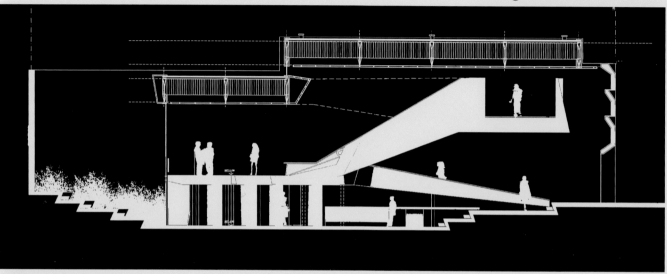

02

【坐落地点】阿根廷 布宜诺斯艾利斯； 【面积】400 m²
【设计】Dieguez Fridman Arquitectos & Asociados 〔阿〕
【灯光设计】Pablo Pizarro 【景观建造】Cora Burgin
【摄影】Juan Hitters

本案商店是一个承载了某种都市文化精髓、移步换景式的展陈空间。在这个空间里，设计师尝试在一个最普通的长方体空间内将多种元素折叠结合，创造了一种非凡的感官体验，成功造就了一个商业和艺术并存的空间。

商店室内空间重塑，打破了常规的概念，给人一种行走在城市空间的错觉。空间的各个元素都显示出一种对比和共生的关系。整个设计还突破了传统的室内透视法，带来了一种无与伦比的探索性质的新鲜购物体验。虽然设计手法不同寻常，但是却不影响空间的功能性。在这里你可以体会到这个空间设计的要义，即创造和衔接所有功能空间的潜在关联，并注重整体效果。

整个空间的色调明亮欢快，材料和灯光的使用和设计都恰到好处。从一层开始，空间便在折叠、弯曲中展开，地面、墙体和顶面在不断衍生的对角斜面中渐次呈现。材料的交替使用和几何学元素在这个空间内部相互作用，线与线的交错，面和面材料上的平滑和肌理的对比，空间中直角和斜面的交替互换。地面由不同的台阶和坡道组成，高低转换。试衣间的造型更像是一个镜面迷宫，在那里你可以从各个角度看到自己。

Store is a carrying case, a kind of urban culture and the essence of the exhibition space. In this space, designers try one of the most common variety of elements within the rectangular space for folding combined to create an extraordinary sensory experience, successfully created a space that contains business and the arts.

The store has been re-shaped interior space, breaking the conventional concept of a sense of the illusion of walking in urban space. The various elements of space shows a contrast and symbiotic relationship. The entire design has also broken through the traditional method of indoor perspective, bringing a unique exploration of the nature of the fresh shopping experience. While the design of an unusual approach, but it did not affect the functionality of space. Here you can experience the essence of the design in this space, namely the creation and convergence of all the features of the potential relevance of space and focus on its overall effectiveness.

The entire space, bright and cheerful colors, materials and the use of lighting and design are just right. Starting from the floor, space will be in the fold, bend, expand, floors, walls and top continuous derivatives in the diagonal slant gradually rendered. Alternate use of materials and geometry elements in this space within the interaction of staggered line and line, surface and surface materials of the smooth and texture contrast, space in the right angle and the slant alternating exchange. On the ground by different composition of steps and ramps, high and low conversion. Shape is more like a dressing room mirror maze, where you can see ourselves from all angles.

↑ 镀锌的外表皮彰显这家商店唯一个性 / Galvanized outer skin highlight the store unique personality.

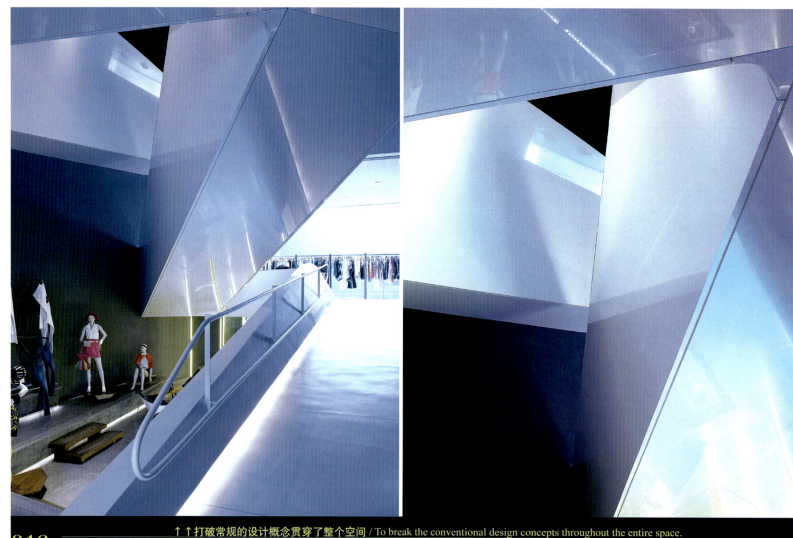

↑↑ 打破常规的设计概念贯穿了整个空间 / To break the conventional design concepts throughout the entire space.
→ 多角度的呈现了这个打破常规的设计概念 / Multi-angle presentation of this design concept to break the routine.
↓ 试衣间的造型像是一个镜面迷宫 / Fitting room shape like a mirror maze; ↓ 楼梯通道 / Staircases

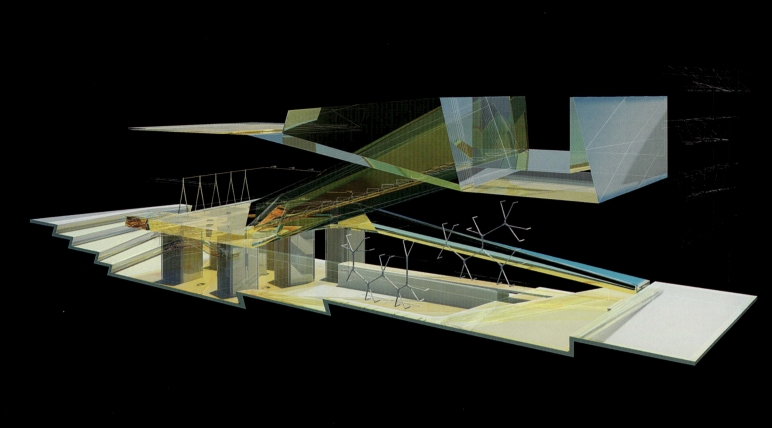

↑↓ 剖面图/ Section
→室外面积不大，堆叠而成的台阶是设计概念的延伸 / Outdoor area is small, the steps of stacked design concept is an extension of.

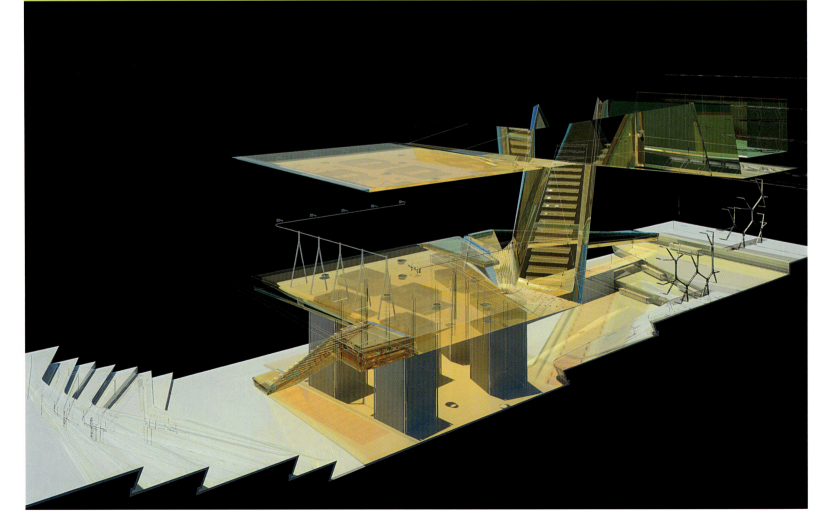

ASOBIO
FLAGSHIP STORE
Asobio 上海旗舰店

03

【坐落地点】上海中山公园 ；【面积】320 m²
【设计】Oki Sato
【设计公司】nendo
【摄影】Daici Ano

设计师精心为ASOBIO进行风格定位，决定用"图片的工作室"作为设计概念的主旨，每个店铺选择一个主题，但都围绕着电影拍摄器材和照相设备展开。

ASOBIO上海旗舰店"龙之梦"的主题是"自我的反射"，或者也可以解释为"镜子"，设计师用"镜子"创造出一个虚幻的世界。店铺入口的橱窗中，几个玻璃镜框高高低低悬挂在背景板上，透过镜框可以看见里面的服饰、货架、天花板，不同的人从不同的角度可以得到完全不同的画面，极富戏剧的感染力。店铺里有两种镜子：一种是"真实的镜子"，方便顾客试穿衣服；另一种是"虚拟的镜子"，像镜子一样的门，人们可以自由通过。黑白灰的色调，使空间中具有了相似性，"镜子"大大小小地穿梭在空间中，同样色彩、材质的框，同样是镶嵌在墙壁上，每一个似乎都是门，又似乎都是镜子，让人无法分辨真假，镜框与门框之间的墙壁被充分利用起来，自然地划分出男装区、女装区及配饰区等，陈列的走向十分清晰，因而空间并不显得拥挤。"镜子"的使用，从视觉上打开了空间的尺度，各个组块之间串联起来，产生有机的联系。

Designers carefully carried out for the ASOBIO style position, decided to use "studio picture" as the main thrust of the design concept, each store to choose a theme, but they all revolved around the film started shooting equipment and photographic equipment.

ASOBIO Shanghai flagship store "Dragon's Dream" the theme of "self-reflection", or it can be interpreted as "mirror" the designers of "mirror" to create an illusory world. Shop windows in the entrance, a few jagged glass picture frame hanging in the background board, can be seen through the frame inside the clothing, shelves, ceilings, different people from different angles can be a completely different picture, very dramatic infectious. There are two kinds of shop mirror: One is the "true mirror" to help customers try on clothes; the other is a "virtual mirror", like the mirror-like doors, people can pass through. Gray black and white colors, so that space has a similar nature, "mirror" large and small, to shuttle in space, the same color and material of the box, also embedded in the wall, each one seems to have the door, it seems that all mirror, people can not tell what is real or false. The wall between the frame and the frame is fully utilized, naturally divided into a men's area, women's area and accessories areas, the display direction is very clear, so space is not cramped. "Mirror" the use of open space from the visual scale, between the various blocks together, produce an organic link.

↑ 自我的反射 / Self-reflection.

↑ 门面全景，对称中略有不同 / Facade Panorama, symmetry in the slightly different.
← 镜子还是门 / Mirror or a door.
↓ 平面图 / Plan

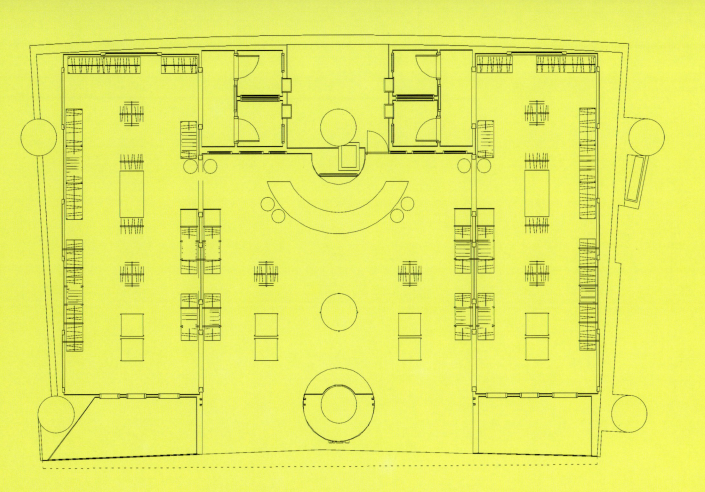

↑ 大画框套小画框 / Small frame large frame sets.
← 镜子造成的对称 / Caused by the mirror symmetry.
↓ 聚光灯经过艺术化的设计 / After the spotlight art design.

DESIGN OF JUSTGOLD
香港镇金店

04

【坐落地点】香港中环国际金融中心商场； 【面积】36 m²
【设计】何宗宪； 【参与设计】Daphne Ng
【设计公司】Joey Ho Design
【摄影】Ray Lau

店铺在空间规划上颠覆了传统金饰店的"面对面"局面，而是模拟艺术馆或画廊的氛围，让客人在自由的空间里浏览货品。设计师摒弃传统的饰柜和单向的展示模式，改为以直立式的壁架和首饰盒般的独立玻璃饰柜来陈列，令首饰变成了令人目不暇接的艺术品。此外，最重视的橱窗部分，亦赋予它创新的圆形饰柜造型，衬托商店弯曲的轮廓之余，有效提高店铺的透明感。

整体造型方面，从女性的妩媚身段取得灵感，以优雅的弧形线条勾勒出全店的轮廓。圆筒形的橱窗饰柜、弧形的壁架造型、不规则的独立饰柜都为店铺带来一抹轻快、跳脱的感觉，设计师更在墙壁上镂空了一条金色缎带，印有镇金店独有的花押图案，衬托背后灯槽，形成一道优雅夺目而且熠熠生辉的环带，与陈列的金饰相互辉映。

发光缎带的尽头，没有摆放传统的收银柜台，引入了一个全新的"Jewelry Bar"概念，模拟酒吧桌子将各款名贵首饰镶嵌其中，再配以高脚凳，让客人一边谈话一边欣赏货品，营造轻松愉快的购物气氛。亮白清新的环境、优雅的线条和崭新的陈列模式，令镇金店成功摆脱传统金饰店的沉闷格局，再次突显品牌领导潮流的地位。

Store in space planning to subvert the traditional goldsmith shop "face to face" situation, but rather simulate the Museum of Art or the Art Gallery of the atmosphere, so that guests in a free space browsing products. Designers to abandon the traditional ornaments display cabinets and one-way model, to be based on vertical wall racks and jewelry boxes as the independence of the glass showcase to display, so that jewelry has become dizzying works of art. In addition, the most valued part of the window, but also give it the showcase of innovative circular shape, set off the contours of the store bent over, effectively improve the transparency of a sense of shops.

The overall shape, the female body made from the inspiration to elegant curved lines sketched out the contours of the shop. Cylinder-shaped windows showcase, curved ledge shape, irregular independent showcase for the shops to bring both light and lively feel, the designer is more hollow in the wall of a gold ribbon, printed with jindian independent some patterns, set off behind the light trough, forming an elegant and eye-catching and shiny ring belt, and the display of gold ornaments embraced.

LED ribbon at the end, there is no display of traditional checkout counters, the introduction of a new "Jewelry Bar" concept, analog bar table, mosaic in which the sections of the precious jewelry, adding a high-legged stool, so that while the guests enjoy the conversation while goods, and create a pleasant shopping atmosphere. Brightening and refreshing environment, elegant lines and a brand new display model, so that the success of the town of gold shop away from the traditional pattern of dull gold shop, once again highlighted the trend of brand leadership position.

↑ 优雅的弧形线条勾勒出全店的轮廓 / An elegant curved lines sketched out the contours of the shop.

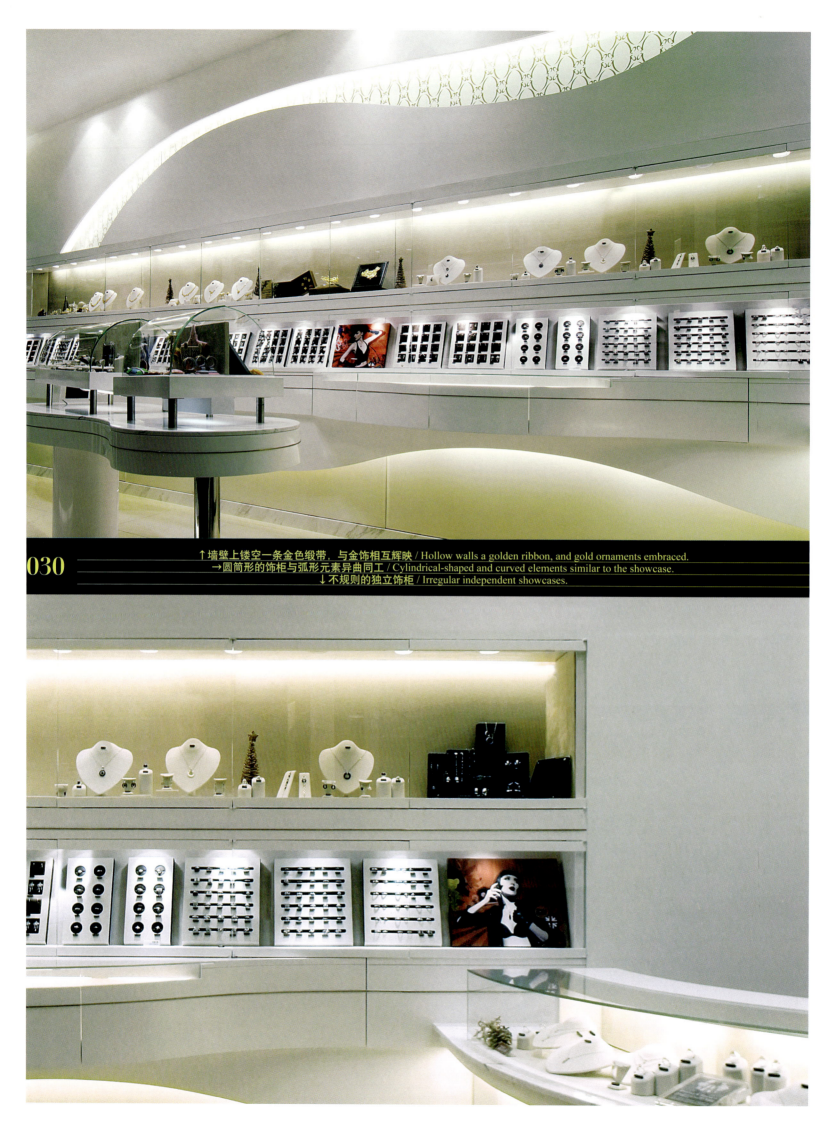

↑ 墙壁上镂空一条金色缎带，与金饰相互辉映 / Hollow walls a golden ribbon, and gold ornaments embraced.
→ 圆筒形的饰柜与弧形元素异曲同工 / Cylindrical-shaped and curved elements similar to the showcase.
↓ 不规则的独立饰柜 / Irregular independent showcases.

↑ 货架与梳妆镜合二为一 / Shelves combined with the mirror.
← 圆形的展示架丰富了橱窗 / Enriched the circular display window.

HCG FLAGSHIP IN GUANGZHOU
和成卫浴广州旗舰店

05

【坐落地点】广州天河区天河北路 ；【面积】380 m²
【设计】李玮珉；【设计公司】李玮珉建筑师事务所
【主要材料】不锈钢拉丝面镀黑板、网点玻璃、LED屏幕墙、米色大理石、中国黑烧面石材、钢构
【摄影】刘涛

步入旗舰店，就翻开了此书的扉页。入口处的水域清澈，将每个刚踏入的观者引入静谧，凝神静气间细细浏览，静静品味，徜徉卫浴的历史长卷；穿行在现代中国卫浴的发展长廊之中。

设计利用不锈钢拉丝面镀黑板的隔断，将产品按照系列分类展示，排布陈列于店中，分门别类如同图书馆的检索表一般清晰。站在店中间只需环顾一周，便将所有产品一目尽览。展柜一面的网点玻璃让灯光透过展柜，形成变化莫测的光影变化。站在大厅中央，沿着米色大理石地铺，和星星点点的蓝色地灯构筑的时光隧道，和成主要产品系列一览无余。

在陈列柜中央，是一个可以休憩的区域平台，除了仓库式展示之外，还划分出了精品生活方式样板展示区，用来陈列创新的、获奖的卫浴套件设计，为消费者带来最新、最时尚的卫浴设计理念。中国黑烧面石材、钢构、冲孔钢板构建的展示空间，利用别致的灯光设计，富有质感的色彩和装饰材料将光洁的洁具衬托得光彩夺目，将一个个时尚、高贵的卫浴空间展示给了观者。极具感染力的影像设计也为该店面增添了耀眼的光彩，二楼外立面的巨大LED幕墙能不断变换光影效果，为每一位来访者提供一场视觉盛宴。

Into the flagship store, opened a book on the title page. At the entrance of water is very clear and each had just stepped into the viewer into quiet concentration, static air between the thin view, quiet taste and feel the history of sanitary scroll; walking in the corridor of modern China's development into bathroom.

Designed to use surface-plated stainless steel wire drawing blackboard partition, have their products classified according to family shows and exclusive fabrics on display in the store, classification is very clear. Looking around the shop just a week standing in the middle of the idea of giving all products a head overlooking. Light through the Showcase, the formation of the vagaries of light and shadow change. Center of the hall stood along the beige marble ground floor shops, and a trickle of blue lamp constructed a time tunnel, the company unfolded all the main product line.

In the showcase central region is an open platform, in addition to warehouse-style show, he also carved out a fine lifestyle model display area is used to display innovative, award-winning bathroom suite designed for consumers to bring the latest, The most stylish bathroom design ideas, in particular, recommend to consumers. Chinese black burn surface stone, steel, punching steel built exhibition space, the use of unique lighting designs, rich textures and colors and decoration materials, sanitary ware will be glossy background was dazzling, will one by one fashion, elegant bathroom space for display to the of the viewer. Highly infectious for the store image design adds a brilliant luster, on the second floor facade of the huge LED wall can be constantly changing lighting effects, for every visitor to provide a visual feast.

↑入口处的水域清澈宁静 / Clear, calm waters at the entrance.

↑ 一条卫浴历史的时空隧道 / A sanitary history of time and space tunnel.
← 绚丽时尚的卫浴展示空间 / Gorgeous stylish bathroom exhibition space.
↓ 一层平面图 / 1st floor plan.

↑ 水景 / Waterscape; ↑ 步入二层的楼梯 / Into the second floor staircase.
→ 灯光透过不锈钢拉丝面镀黑板隔断 / Light through the stainless steel surface coated wire drawing cut off the blackboard.

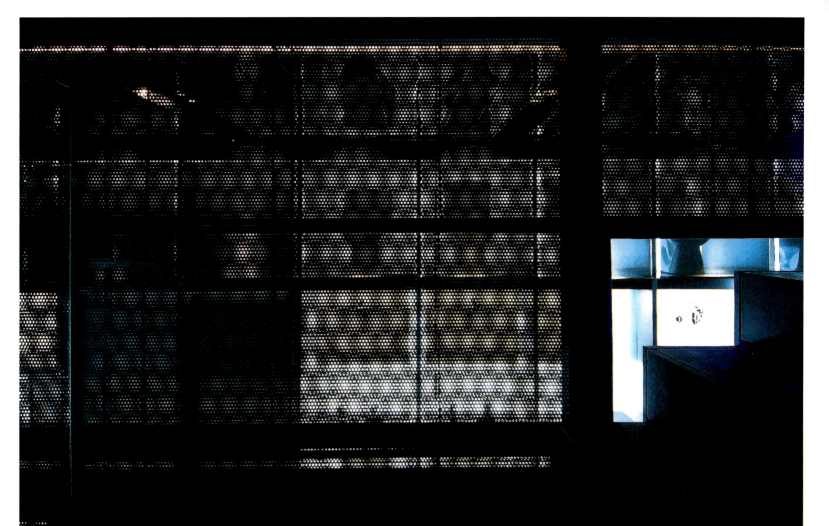

↑不锈钢拉丝面镀黑板的隔断 / Surface-plated stainless steel wire drawing blackboard cut off.
←别致的灯管设计凸显了产品的个性 / Chic lamp design highlights the product's personality.
↓二层平面图 / 2nd floor plan.

GALERIES
LAFAYETTE SHOP
Galeries Lafayette店铺

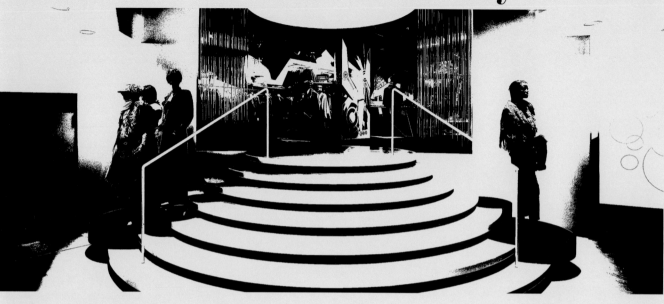

06

【坐落地点】德国柏林
【面积】502 m²
【设计】plajer & franz Studio 〔德〕
【摄影】diephotodesigner.de

店铺所在的建筑比较特别，设计师把这个空间定义为一个小型的展陈空间，希望能创作出一个与建筑的身份相符合的、足够别致的地方，通过灯光、色彩、空间划分等各种设计手法，以及对高度灵活的展览概念的运用，让这里更加具有吸引力，从而也能够更好地体现出品牌的内涵。

一切的内部空间展开的源头则来自中间的那个圆锥体，五个环形从上到下依次环绕，自然形成一个多层次的吊顶和三个陈列架，材质的镜面效果和灯光的配合交相辉映。这个设计利用环形的变化和灯光的相互作用，让室内空间焕然一新的同时又不破坏原建筑空间所呈现的独一无二的多角度的特质。

入口处的楼梯由七个圆形叠加组合而成，其中有五个黑色的圆台从这里衍生出来，用于摆放模特。在这样的场所中，曲线的运用总能够将其打造的更加典雅柔和，让来到这里的顾客流连忘返；室内的墙面上也用墙纸等材料拼合了各种线、面的圆形，呼应整个的设计概念；试衣间是圆柱形的，有着灵活可变的门可以自由的开阖。既要让空间显得格调高雅，又要突出商品的品质，在这一点上设计师让灯光明亮而不炫目，很好地营造出了这里轻快而不失高贵的购物氛围。

Comparison shop building where the special designers to this space is defined as a small exhibition space, hoping to create a consistent identity and architecture, adequate chic place. Shop by lighting, color, space division and other design techniques, as well as highly flexible use of the concept of the exhibition, and make it more attractive, and thus better able to reflect the brand connotations.

Expand all sources of interior space comes from the middle of that cone, five ring from top to bottom in turn surrounded by the natural formation of a multi-layered ceiling, and three display racks, material and lighting with the mirror effect to perfection. This design change and the use of circular light interaction, so that interior space a new look at the same time without destroying the original building space presented by multi-angle characteristics.

A circular staircase at the entrance by the superposition of a combination of seven, including five black circular platform derived from here, for the display model. In such establishments, the use of the curve can always be more elegant to create soft, so that customers come here to visit; interior walls with wallpaper and other materials are also put together a variety of lines, circular surface, corresponding to the entire The design concept; dressing room is cylindrical, has a flexible variable-door can be freely switch. We should let the space look elegant, but also highlight the quality of goods, at this point instead of the designer so that bright lights dazzling, well here to create a light and elegant shopping atmosphere.

↑ 五个环形从上到下依次环绕 / 5 from top to bottom turn around the ring.

↑入口处的楼梯由七个圆形叠加组合而成 / Circular staircase at the entrance by the superposition of a combination of seven.
←一个多层次的吊顶和三个陈列架 / A multi-layered ceiling, and three display racks.
↓墙面的设计也同样出色 / The wall design is equally impressive.

↑↑ 多种材质的运用 / The use of a variety of materials.
← 材质的镜面效果和灯光的配合 / Material with the mirror effect and lighting.
↓ 平面图 / Plan

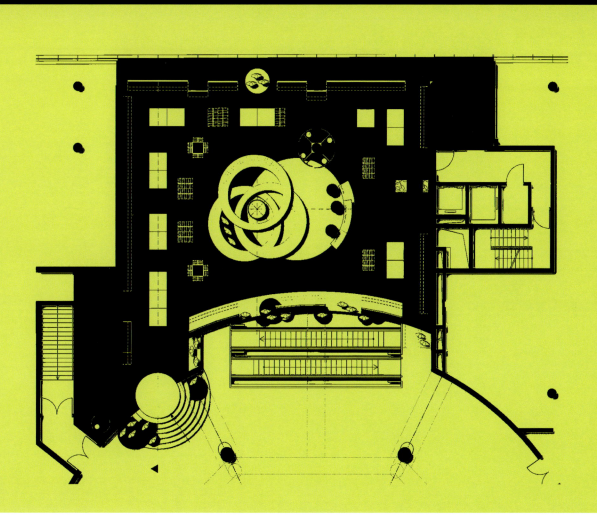

INTERIOR DESIGN OF KIKI 2 SHOP
Kiki 2专卖店

07

【坐落地点】荷兰马斯垂克市；【面积】34 m²
【设计】Maurice Mentjens 〔荷〕
【主要材料】橡木饰面板、环氧基树脂地板、卤素聚光灯、玻璃、镜面材质、中密度纤维板
【摄影】Arjen Schmitz

本店是Kiki Niesten的一个经销分店，店主就希望这个分销店铺的室内设计可以在集功能、时尚、奢华于一身体现Kiki Niesten精神实质的同时，又不乏风趣幽默的元素。Kiki 2的设计师认为这个店面整体概念就像一个看上去有些古老和陈旧的储藏室，因此包括墙面和天花板在内，几乎所有的室内空间都被灰色的、橡木材质的、高低错落的货架和小橱柜充斥着。

店内的光源被巧妙地隐藏在天花板上层次分明的小隔断中，而柜台之后的矩形空间内，柔和明亮的灯光自上而下投射于地面。通向上层的门，就掩藏在店内柜台的左手边上。邻近于柜台的巨大的玻璃墙货架，如同柜台后的一个边界，当前门打开时，它可以避免直接承载所有的重量。店铺的尽头是两个挂衣间，它们与和柜台同一面的橱柜相对应，那里有三个不同的档可以挂衣服。所有的搁板都经过精心设计和计算，可以放置在任意的衣橱中。

设计师通过独特材质的组合，朴素而实用的设计手法成功打造了一个个性化的购物空间。整个设计在正与负、体量与空间、凸与凹之中体现出别样的世俗、安定、平凡的本质，表露出一种空间、材料在与光线的对舞之后，所呈现的一种诗情的轻盈。相信定能给来此挑选物品的顾客带来别样的感受。

Kiki Niesten restaurant is a branch of a distribution, shopkeepers on the hope that the distribution of shops in the set features interior design, fashion, luxury in a spiritual essence embodied Kiki Niesten the same time, but also no shortage of humorous elements. Kiki 2 designers think that the whole concept store looks like a somewhat archaic and obsolete storage room, so including the walls and ceilings, including almost all of the interior spaces have been gray, oak material, the high and low scattered in shelves and small cupboard filled.

The store's lighting has been cleverly hidden in the ceiling of a small layered cut off, while the counter after the rectangular space, bright lights soft top-down projection on the ground. Leading to the upper door, hidden in the shop on the left-hand side on the counter. Adjacent to counter a huge glass wall shelves, like behind the counter of a boundary, the current door is open, it can avoid a direct bearing all the weight. The two shops are the two hanging clothes between them and the counter corresponding to the same side of the cupboard, where there are three different files that you can hang clothes. All the shelves have been carefully designed and calculated, can be placed in any closet.

Designer through a unique combination of materials, simple and practical design techniques to create the success of a personalized shopping space. The entire design in positive and negative, body mass and space, convex and concave is reflected in a different kind of secular, stability, and extraordinary nature, revealed the kind of space, materials contrast with the light, after a poem presented by love the light. I believe we can give to the selection of items to bring customers a special feeling.

↑ 店铺的整体概念就像一个陈旧的储藏室 / The entire concept of a shop like the old storage room.

↑ 展示架 / Display
← 地上纹样像丢弃着的白色旧衣服 / Patterns on the floor like a discarded old clothes with white.
↓ 平面图 / Plan

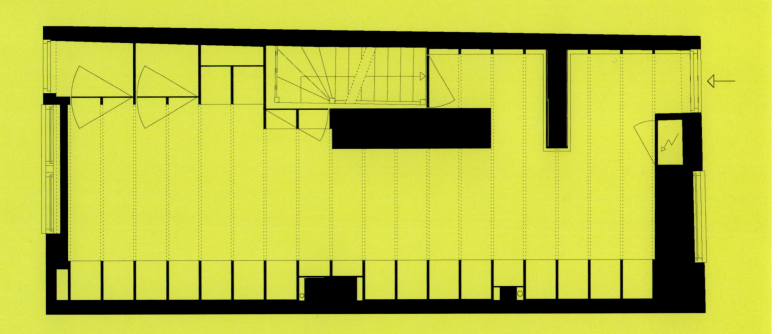

←↑ 几乎所有的室内空间都被高低错落的灰色橡木材质货架和小橱柜充斥 /
Almost all of the interior space are scattered high and low gray oak shelves and small cabinets full of material.

↓ 立面图 / Elevation

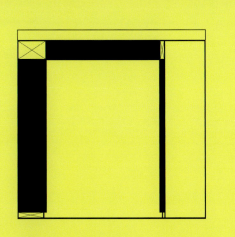
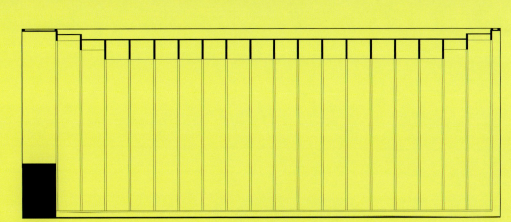
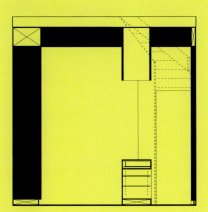
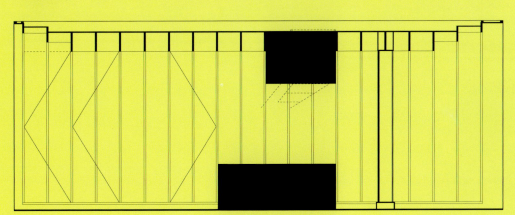

A MEN'S MONOPOLIZED STORE
Geometry男装专卖店

08

【坐落地点】德国 柏林
【面积】100 m²
【设计】plajer & franz studio〔德〕
【摄影】Ken Schluchtmann

设计师决心要设计一个全新的"男人世界"。专卖店有一个口号:"我们能有多不可思议!"设计师在保证使用功能的前提下,努力让这个作品看起来与众不同。令人惊奇的是,当你第一眼看见这个专卖店的时候,恍惚中像走进了一个老教授的家,那种井井有条的景象甚至能让你推断这肯定是个数学教授的家。

这一定是个有条理的教授,并且擅长分析比例关系和做分类整理。店铺里面有很多分类放置的商品,像相册一样分门别类,有条不紊,仿佛在静静陪伴它的主人沉迷于数学的海洋,思考、计算、分析。这里所有对称的设计(货架)和不对称的设计(呈角度放置的青铜镜子形成的反射效果)都和谐相处,灰色的墙壁、黑色的木地板和白色的喷漆橡木家具又都是流行的设计元素。

为了营造一个独一无二的氛围,设计师运用了许多家庭式的元素,例如等待区的小地毯和人造卫星造型的吊灯。而店铺内部真正的空间设计还是以实用性为主,充分考虑了顾客的购物习惯,方便人们挑选。

Designers are determined to design a new "man in the world." Store has a slogan: "We can have incredible!" Designers use function under the premise of ensuring efforts to make this work looks different. The surprising thing is, when you see this store at first sight, when in trance, like an old professor walked into the home, the kind of coherent vision or even allow you to infer that this is definitely a math professor's house.

This must be a structured professor, and the analysis of the proportion of good relationships and do sort out. Shops there are many categories of goods to be placed, like the album, like categorized, organized, as if to accompany its owner quietly indulge in mathematics ocean, thinking, calculation and analysis. Where all the design symmetrical and asymmetrical designs are in harmony, gray walls, dark wood floors and white painted oak furniture, and both are popular design elements.

In order to create a unique atmosphere, designers use a lot of family-style elements, such as waiting area rugs and satellite shape of chandeliers. Inside the shop or in the real space design practical-based, fully taking into account customer's shopping habits, easy for people to choose.

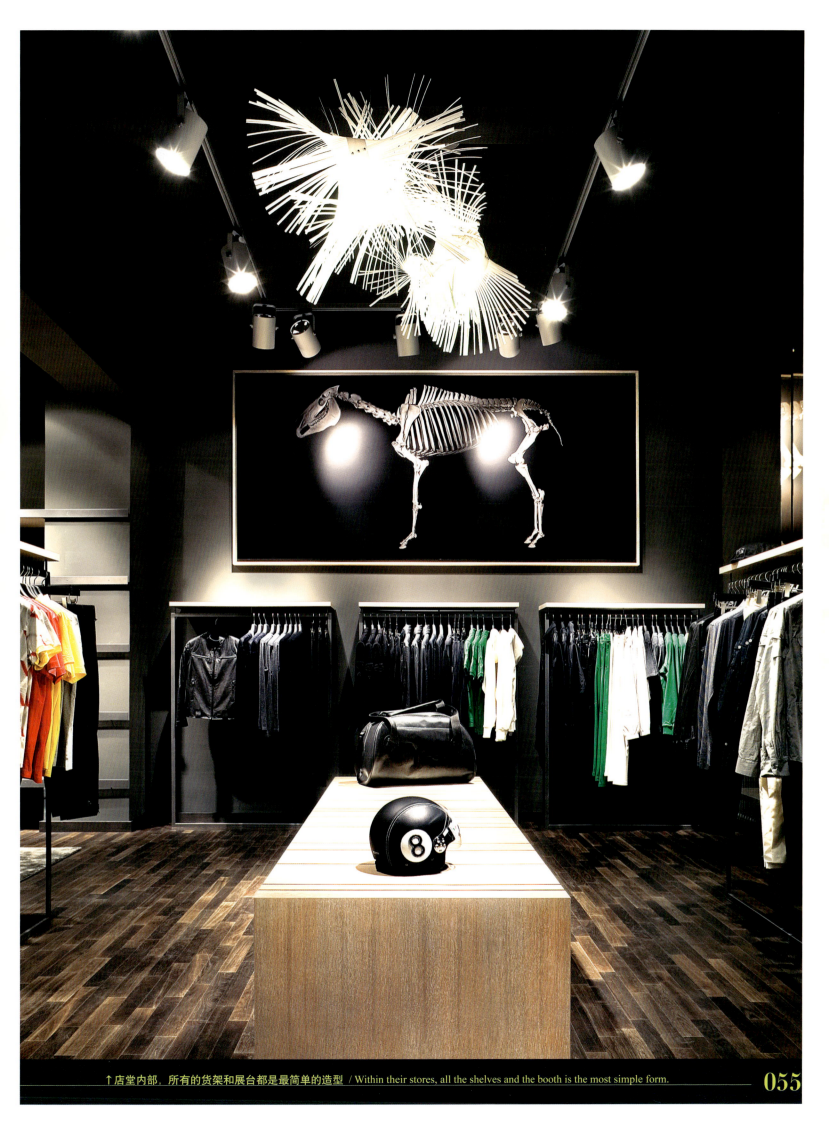

↑店堂内部，所有的货架和展台都是最简单的造型 / Within their stores, all the shelves and the booth is the most simple form.

↑店铺外观 / Store appearance.
→没有过多装饰的橱窗，以直线条为主 / Without too many decorative windows, the main section in a straight line.
↓店堂内部的卖品分门别类，很有条理 / Their stores for sale within the different categories, it is structured.

↑↑试衣间和天花都选用了恐龙的图案 / Fitting room and smallpox have chosen dinosaur pattern.
←等待区，舒适的座椅和人造卫星造型的吊灯 / Waiting areas, comfortable seating and satellite-shaped chandelier.
↓平面图 / Plan

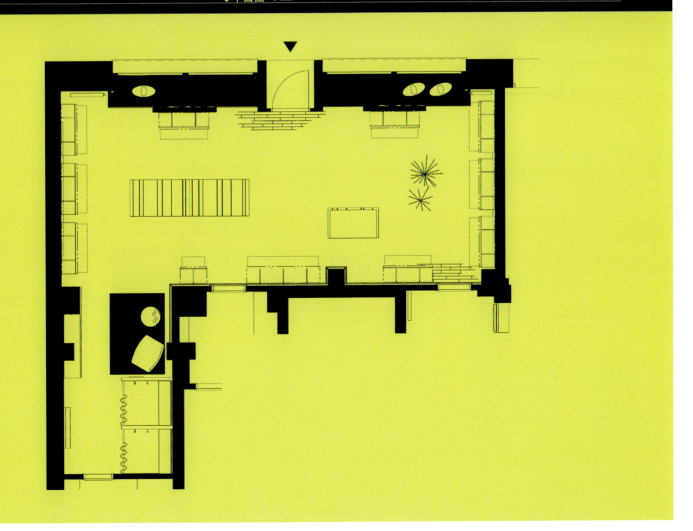

ANN
DEMEULEMEESTER'S SHOP
韩国某服饰店

09

【坐落地点】韩国 首尔
【面积】735 m²
【设计】Minsuk Cho〔韩〕
【设计公司】Mass Studies〔韩〕

建筑三面郁郁葱葱的绿色正是整个建筑的序曲，这个结合了Rock、青春和军装风格的服饰品牌。用常绿植物作为建筑外表面，用绿色的植物来配合她多变的面料和剪裁。

设计师通过一个如何用弯曲的混凝土外墙给植物安家的设计思路，用带点解构主义的设计手法去协调这个19世纪比利时风格建筑所特有的高窗户。它能提供一个微循环生态系统，保证富贵草的生长需要，建筑的外表还动用了地热装置，以维持整个绿色外衣的生态需要。设计师把整个建筑作为一个自然系统来考虑，避免城市建筑普遍存在的低层高，高密度式的设计模式。建筑的设计上充分考虑到自然与人工，室内与室外的融合关系，而不是让它们对立起来。

一楼室内就是这个著名的服饰品牌，而它内部的天花做成了一个起伏的巨大的波浪形。内部有钢管做必要的结构支撑。这样的设计形成了一个弓形的开口，使店铺能最大程度地面对街道和外面的竹篱笆。通过楼梯进入餐厅的主入口，楼梯是建筑的东边最重要的部分。二层的餐厅空间受到了一层店铺波浪形的天花影响。而通往地下层的楼梯入口是一个白色的狭长的通道，根据建筑结构的变化越来越开阔，慢慢形成了一个宽阔的视野。地下层的感觉像一个布满了苔藓的山洞。

Three sides by lush green building is a prelude to the whole building, the combination of Rock, youth and uniform-style clothing brand. With the outer surface of evergreen plants as construction, using green plants to cope with her ever-changing fabric and tailoring.

Designer through a curved concrete wall how to settle down to the plant design ideas, with a little deconstruction of the design techniques to co-ordinate the 19th century style of architecture unique to Belgium, the high windows. It provides a micro-cycle of the ecosystem to ensure the growth of rich grass needs, building exterior is also used in geothermal installations, in order to maintain the ecological needs of the entire green coat. Designer the whole construction as a natural system to take into account to avoid the urban architecture of high prevalence of low-level, high-density style of design patterns. Architectural design of fully take into account the natural and artificial, indoor and outdoor integration relationship, rather than let them against each other.

On the first floor room is this well-known clothing brands, and its internal undulating ceiling made of a huge wave-shaped. Within the pipe to make the necessary structural support. This design creates a bow-shaped opening, so that premises can be to maximize the face of the streets and outside the bamboo fence. The stairs into the restaurant through the main entrance, construction of the east staircase of the most important part. The second floor restaurant space is a layer of wavy ceiling effects shop. The staircase leading to the basement entrance is a white narrow channel, according to architectural changes in the structure more open, and slowly form a wide field of vision. Basement feels like a cave covered with moss.

↑ 建筑外观 / Architectural appearance.

↑ 楼梯的入口处是一个狭长的白色空间 / Staircase at the entrance is a narrow white space.
↓ 从走廊向外可以看到建筑极具特色的绿色 / Can be seen from the corridor outside the green building very unique.

↑ 从楼上俯瞰小天井 / Upstairs overlooking a small courtyard.
↓ 一层平面 / 1st floor plan；↓ 二层平面 / 2nd floor plan.

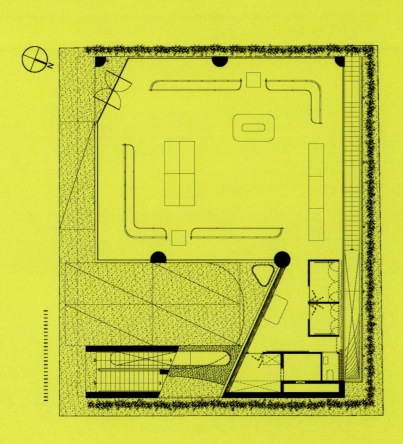
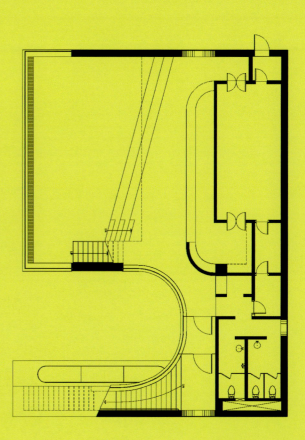

↑ 二楼的餐厅区 / On the second floor of the restaurant area.
→ 狭长的楼梯 / Narrow staircase.
↓ 三层平面 / 3rd floor plan；↓ 顶层平面 / Top floor plan.

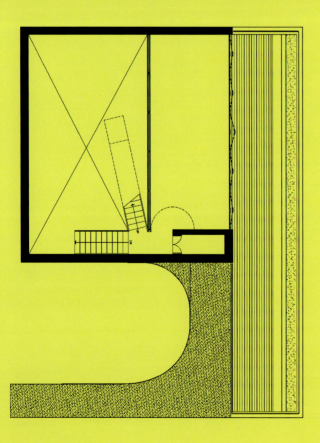
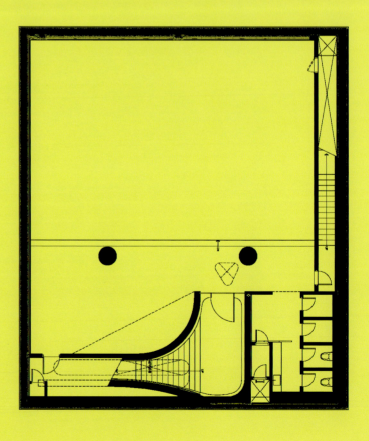

↑ 从楼梯向上 / From the stairs up.

↓ 店铺内景 / Shop interior.

↑↓ 店铺内景，起伏的天花很有震撼力 /
Shop interior, undulating ceiling is very shocking.

JEAN-PATOU
PERFUME SHOP

Jean-Patou香水店

10

【坐落地点】巴黎，法国
【面积】150 m²
【设计】Eric Gizard 〔法〕
【主要用材】水晶玻璃、有机玻璃、毛毡、射灯

香水店坐落于一座有着拱廊的传统经典欧式建筑之中。设计师Eric Gizard充分考虑了这个品牌优雅、简洁的品质，使用了简约现代的装饰风格，融合了部分的透明结构，从室内设计的方面对整个空间的语言进行了重新演绎，并在其橱窗使用了40 mm厚的水晶透明材料，突出了空间独特的质感。

底楼的店铺内使用了一面毛毡质地的墙壁，并配有8个圆柱结构，呈现不同产品的风格。在圆柱展台后那一块粉红色的玻璃，突现了创始人对此色彩的偏好，圣地弗里亚粉红及圆柱展台和墙壁的特定色泽交相辉映。琥珀色的香水在透明的树脂块体的衬托下更显剔透晶莹。冷暖色调对比的娴熟运用，让设计在隐约中再现永恒的创新意识及层次感和神秘感。这些装饰协同不时变化颜色的柔和灯光，尽显了产品的优越品质。悬在圆孔中的香水瓶更是宛若天空中的精灵，吸引着顾客的注意力。穿插其中的粉色在柔情中体现了精致，将典雅的风格融入于生活情趣之中，使进入商店的人们忘却了外面世界的俗世纷扰。

Perfume shop is located in an arcade with classic European architecture into the traditional. Designer Eric Gizard full consideration to the brand elegant, simple quality, using the simple modern style, blending some of the transparent structure, from the interior design aspect of the language of the entire space has been re-interpreted, and used in its window 40 mm thick crystal transparent material, highlighting the unique texture of space.

The use of ground floor shop side of the felt texture of the walls, and is equipped with eight cylindrical structure, and create a different product styles. After the cylinder was a booth in the pink glass, emergence of the founders of this color preferences, the Holy Land Fria pink and cylinder stand and the walls echo the specific color. Amber perfume in a transparent resin block the backdrop of even more crystal clear. In contrast to the use of warm and cold colors, so that the reproduction of subtle design in the eternal sense of innovation and levels of influenza and mystery. These decorations together from time to time change the color of the soft lights, filling the superior quality of the product. Suspended in the hole in the perfume bottle is like a wizard in the sky, attracting the attention of customers. Pink interspersed among them reflected in the tenderness of the refined and elegant style will be incorporated into a delight of life among the people who forget to enter the stores of the turbulence of the outside world.

↑ 悬在圆孔中的香水瓶 / Suspended in the hole in the perfume bottle.

↑ 透过透明的橱窗可见商店内部的精致 / Can be seen through the transparent window within the exquisite shops.
↓↓ 透过粉红的玻璃可见开放性的香水吧 / The glass can be seen through a pink perfume open bar.

↑女人们可在此享受高级香水的芬芳 / Women can enjoy high-level perfume fragrance.
↓↓VIP体验区 / VIP Experience Zone.

BASTARD
FLAGSHIP SHOP
Bastard旗舰店

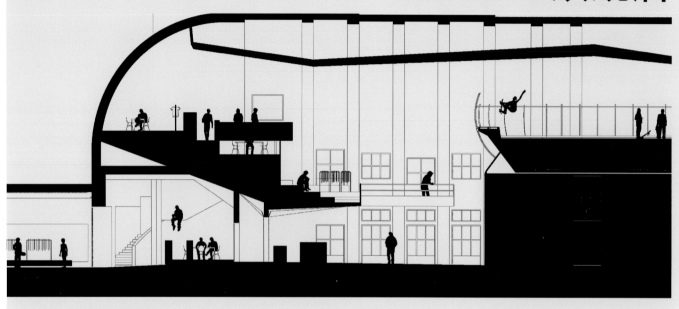

11

【工程名称】意大利Bastard旗舰店
【面积】1400 m²
【设计】Studiometrico〔意〕
【摄影】Giuliano Berarducci

旗舰店位于建筑主入口处,面积70 m²,人们从这里进入Bastard世界。穿过旗舰店就进入了行政部,这里是原电影院的半月形门厅,沿墙壁两侧为两个弧形楼梯通往上层楼座,几扇玻璃窗与底部池座相连。行政部连接所有主要的空间,建筑的主轴线与入口处偏转的轴线在此交汇。原有大理石地面略有倾斜,设计者木制平台界定了新的功能空间,并用木挡板围合了三个半开敞空间,为员工提供私密的办公空间,屏蔽隔壁店铺中顾客好奇的目光。

沿弧形楼梯到达二楼设计部,这里原为电影院楼座,设计部利用了升起的台阶,在不同标高上搭建平台,用钢结构固定到现有楼座的混凝土梁上。平台错落有致,既相互贯通又互不干扰,木板围合出各自独立的工作空间。楼座底部的踏步未作修改,作为开放的展示空间,为弧形楼梯保留一个自由灵活的空间。产品仓库建在15 m高、空间巨大的池座上,此处空间单一放置大量仓库货架,简洁实用,由二层钢结构组成。碗形溜滑板场悬挂于仓库6 m高的上方,碗形溜滑板场与拱形屋顶结构相得益彰。把200 m²的"碗"置于仓库之上,既节省了空间,又与楼座上的设计部相互呼应。二者都位于建筑主轴线上,形成了视觉和空间上的联系。

Flagship store is located in the construction of the main entrance area of 70 m², people from here into the Bastard in the world. Entered through the flagship store of the Administrative Department, here is the original half-moon cinema hall, along the walls on both sides of the two curved staircase leading to the upper balcony, a few glass windows and connected to the bottom of the parterre. Administration Department to connect all the major space, building at the entrance to the main axis with the axis of deflection at this intersection. Slightly tilting the original marble floors, designer wooden platform to define a new function space, and use wood baffle enclosure of the three and a half open space, to provide staff with private office space, shielded by customers in stores next door to the curious eyes.

Along the curved staircase to reach the second floor of the design department, where cinema balcony was originally designed used by the Department of the rising level, in different elevation on a platform with a steel structure fixed to the existing balcony concrete beam. Platform patchwork, both run through each other without disturbing each other again, wood enclosure a separate work space. Product warehouse built in the 15 m high, a huge space on the parterre, where a large number of warehouse space to place a single shelf, concise and practical, composed by a two-story steel structure. Slide skateboard bowl games are hung on the warehouse 6 m high above the bowl slide skateboard field and arched roof structure complement each other. To 200 m² of the "bowl" placed above the warehouse, saving space, but also with the design department on the balcony each other well. Both are located in the building axis line, forming a visual and spatial connections.

↑ 高低错落的设计部 / Scattered high and low of the design department.

↑ 钢结构组成的货架 / Formed steel shelves.
↓ 一层平面图 / 1st floor plan.

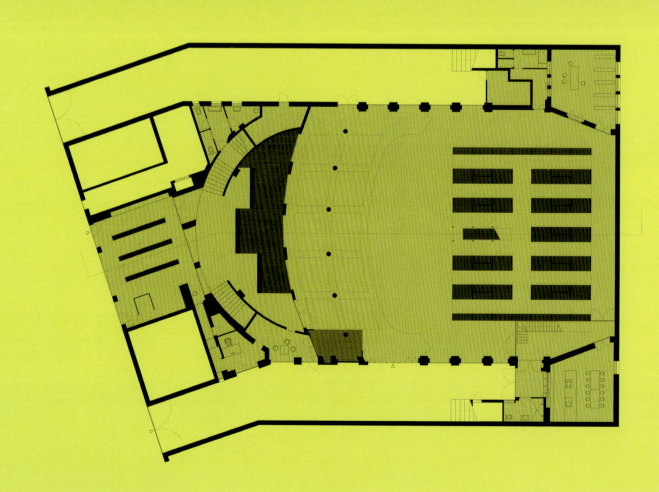

↑ 通向二楼的楼梯 / Staircase leading to the second floor.
↓ 二层平面图 / 2nd floor plan.

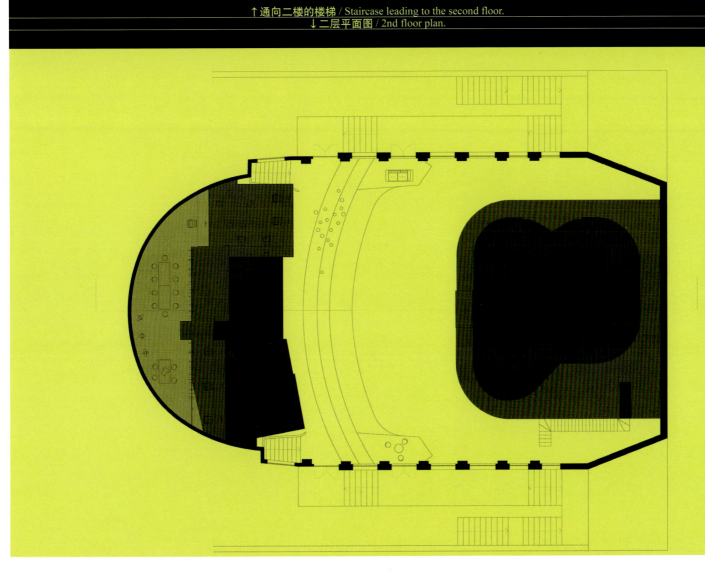

↑ 展示部分显得随性自然，后面的小空间是设计部 / Displayed some appear as naturally, the small space behind the design department.
↓ 设计部 / Design department.

↑ 设计部 / Design department.
↓ 碗形溜滑板场 / Slide skateboard bowl games.

NOVO
SHOPS IN BEIJING
Novo北京服装店

12

【坐落地点】北京朝阳区建国门外大街永安里乙12号LG双子座大厦2楼；【建筑面积】239 m²
【设计】小川训央（NORIO OGAWA）；【参与设计】CHIKARA SASAKI
【主要用材】山纹松木做清漆上色、12 mm的强化玻璃、局部贴磨砂贴膜、局部墙壁贴绒毯
【摄影】Minoru Iwasaki

当有客人前来的时候，一家店才能真正体现出它的价值。商业设施的设计，其实也就是带给前来购物的客人一个舒适的空间，而与此同时，必须要考虑的是，对业主来说，需要的是一个"流行的店"，一个"热卖的店"。

所谓商业设施，就是各店都有自己的营业目的，人们为了满足这个目的汇集而来。作为"新概念SHOP"，novo一直散发出其独有的风格和魅力。近年来，随着在中国各地业务的不断开展，novo的发展势头不见减弱，反倒愈发强劲。novo的两个店铺设计，设计概念却截然不同。

北京双子座大楼店的设计，则以黑灰为色彩基调，配合木格子层板，从整体上展现出既时髦又不失品位的高级感。通过高低差的巧妙配合，尤其是越往店内深处去，地板抬得越高的设计，使人们从店外就能一眼看到店内的最深处，对于店内经营何物自是一目了然。

When guests come, when a store can really demonstrate its value. The design of commercial facilities, in fact, that is, to bring the shopping guests a comfortable space, but at the same time, we must consider is that for property owners, is needed is a "pop shop", a "best-selling shop."

The so-called commercial facilities, that is, each store has its own business purposes, people come together in order to meet this purpose. As a "new concept SHOP", novo has been distributed out of its unique style and charm. In recent years, with operations in China over the ongoing, novo increasingly strong momentum of development. The two shops novo design, design concepts are very different.

Beijing Gemini building design shop, places black ash for color tone, with lattice wood laminates, from the whole show high sense of fashion taste. Through the height difference with the ingenious, particularly in the more deep to go to the store, the floor design of the higher lift, so that people from outside the shop will be able to see inside one of the most deep, for store operations at a glance what things naturally.

↑ 流线型的道具和日光灯的排列 / Streamlined arrangement of props and fluorescent lamps.

083

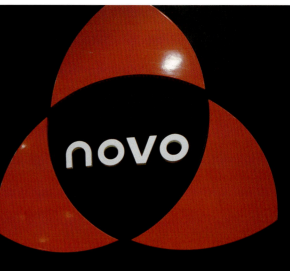

↑ 从入口处看向店铺深处 / From the entrance to the shop to see the depths.
↓ 入口位置的道具，用于陈列新款或经典款商品 / Entrance location props for the display of new products, or the classic models.

↑陈列区域中的沙发设计，为客人提供了悠闲的购物氛围 / Display area of the sofa design, providing guests with a relaxed shopping atmosphere.
↓平面图 / Plan

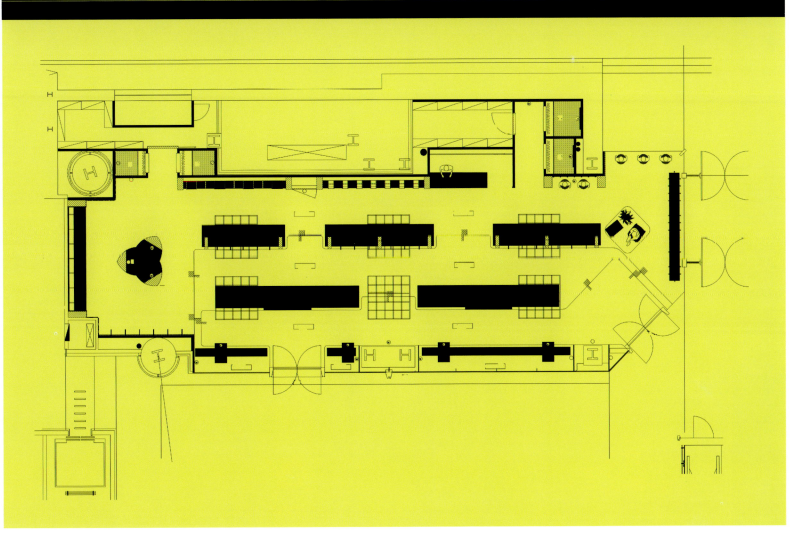

NOVO
SHOPS IN SHANGHAI
Novo上海服装店

13

【坐落地点】上海浦东新区陆家嘴西路168号正大广场1楼；【建筑面积】800 m²
【设计】小川训央（NORIO OGAWA）；【参与设计】CHIKARA SASAKI
【主要用材】油漆、铁管、视觉图案、造型灯具
【摄影】贾方

当有客人前来的时候，一家店才能真正体现出它的价值。商业设施的设计，其实也就是带给前来购物的客人一个舒适的空间，而与此同时，必须要考虑的是，对业主来说，需要的是一个"流行的店"，一个"热卖的店"。

所谓商业设施，就是各店都有自己的营业目的，人们为了满足这个目的汇集而来。作为"新概念SHOP"，novo一直散发出其独有的风格和魅力。近年来，随着在中国各地业务的不断开展，novo的发展势头不见减弱，反倒愈发强劲。novo的两个店铺设计，设计概念却截然不同。

上海正大广场店的设计，是在重视novo原本就有的运动气息的基础上，重点尝试了流线型造型，利用道具及店内装饰来演绎出流畅的速度感。

When guests come, when a store can really demonstrate its value. The design of commercial facilities, in fact, that is, to bring the shopping guests a comfortable space, but at the same time, we must consider is that for property owners, is needed is a "pop shop", a "best-selling shop."

The so-called commercial facilities, that is, each store has its own business purposes, people come together in order to meet this purpose. As a "new concept SHOP", novo has been distributed out of its unique style and charm. In recent years, with operations in China over the ongoing, novo increasingly strong momentum of development. The two shops novo design, design concepts are very different.

Shanghai Super Brand Mall store design is the emphasis on novo There were originally based on the movement of breath, focusing on tried streamlined shape, the use of props and in-store decoration, playing with a smooth sense of speed.

↑ 流线型的道具和日光灯的排列 / Streamlined arrangement of props and fluorescent lamps.

↑ 从入口处看向店铺深处 / From the entrance to the shop to see the depths.
↓ 入口位置的道具，用于陈列新款或经典款商品 / Entrance location props for the display of new products, or the classic models.

↑ 从店铺深处看向入口位置 / From the shops to see the depths to the entry location.
↓ 平面图 / Plan

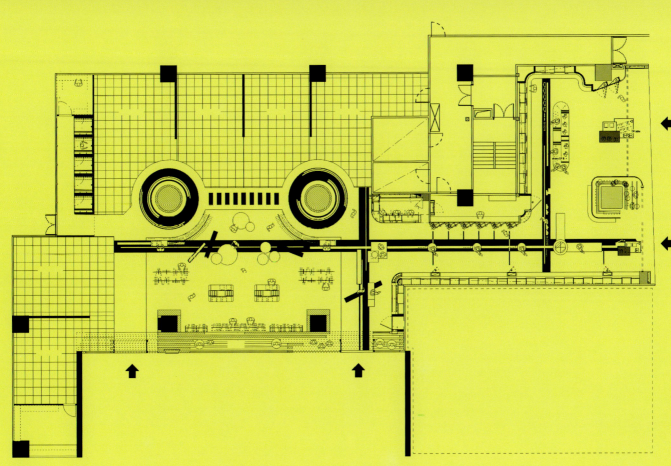

CUSTO SHOP'S DESIGN IN BARCELONA
Custo巴塞罗那店

14

【坐落地点】西班牙巴塞罗那Claudio Coello91号； 【面积】280 m²
【设计】Teresa Sapey建筑事务所
【合作设计】Angela Sanz, Agnisezka Cohojnacka, Raquel Rojas, Marta Melendo
【摄影】Courtesy Teresa Sapey - A.L. Baltanás y Pablo Orcajo

商店内的照明在视觉上扩大了整个展示空间。这里的地面、色彩、墙面和灯效等设计都为商店营造出一种轻松、愉悦的氛围，同时也掩盖并隐藏了空间层高的问题。

从户外很难看到整个零售店的内部，因此也为顾客创造了更多的意想不到。零售店最开始给顾客的是一种冷漠的空间感受，但这种感受很快就被与室外风格完全相反的室内设计所颠覆了，消费者可以在店内找到Custo巴塞罗那系列的男、女装以及配饰。

商店地下室被设计成一个展示空间，一张紫红色的地毯是这里的主要元素，整个设计主要是以形式和色彩共同打造了一种巴塞罗那地区的风格和图案。

Lighting shops in the visual display space as a whole expanded. Here on the ground, color, walls and lighting and other designs for the store to create a relaxed, pleasant atmosphere, but also to cover up and hide the problem of high spatial layers.

From the outside it is difficult to see inside the store and, therefore, for customers to create more unexpected. Retail stores to customers is the beginning of a cold room for feelings, but this feeling will soon be the exact opposite with the outdoor-style interior design upside down, consumers can be found in the shop Custo Barcelona series of men, women as well as a ornaments.

Shop basement is designed as a display space, a purple carpet is a key element here. Mainly in the form of the entire design and color work together to build a regional style and design in Barcelona.

↑ 零售店入口 / Store entrance.

↑ 与室外风格完全相反的室内空间 / And outdoor style of the opposite interior space.
↓ 休息区 / Rest area; ↓ 装饰沙帘 / Decorative curtain.

↑ 店内大面积的玻璃 / Store large areas of glass；↑ 货架 / Shelf
↓ 白色空间 / White space；↓ 灯光设计 / Lighting design.

VILLA MODA
IN BAHRAIN
Villa Moda巴林店

15

【工程名称】Villa Moda 巴林店
【面积】1 050 m²
【设计】Marcel Wanders Studio〔荷〕
【摄影】Courtesy of Marcel Wanders Studio

露天剧院作为巴林的地标景观之一，以其颇具性格的混杂吸引着四方游客。Wanders决定将设计定位为"国际露天剧院"，即保留露天剧院混杂丰富的精神，注入现代的前卫奢侈元素，使之成为一个融古典中东风情和现代国际风尚于一体的设计空间。它更像是一个小城市，通过多种因素的综合，展现出建筑的美感。

Villa Moda入口处的墙面由众多巨大"珍珠"装饰而成，这些球形珍珠已经成了该店的标识之一。此外，长达10m多的入口通道的计灵感亦是来自巴林海湾国家的性质。地域文化通过别具一格的设计被放大展现，大气之中尽显设计师的细腻。珍珠墙和悠长的隧道入口，通向一个未知的奇幻世界，所有的一切都像爱丽丝漫游奇境的前奏。

追求惊奇效果的Wanders，希望店铺给人带来的一种丰富饱满的立体感，创造一种移步异景的应接不暇。进入店铺后，挑高的空间，极简的色彩运用混杂着强烈的中东风，给人极大的视觉冲击。黑白的主色调给人一种端庄克制之感，搭配以大面积复杂的花纹，点缀少量的蓝色玻璃橱窗和黄色墙面，制造一种动静皆宜的感觉，极简与繁复的碰撞。室内的设计重点均用白色表现。其穹顶式构造，呼应了当地伊斯兰的圆顶建筑，又增加了空间的层次感。墙面的白色浮雕大花，不对称的舒展之美，给人一种略带压迫感的震撼。

Open Theater as one of Bahrain's landmark landscape, with its rather mixed character of attraction for tourists. Wanders the decision will be designed as "an international open-air theater," that is a rich mixture to retain the spirit of open-air theater, infusing it with modern avant-garde luxury elements, making it a melting of classical style and modern Middle East, the international fashion in one of the design space. It is more like a small city, through a variety of elements, to show the beauty of architecture.

Villa Moda from the wall at the entrance to a huge number of "pearl" decorations made of these spherical pearls have become one of the store's logo. In addition, up to 10m of the total number of the entrance channel is also inspired by the nature of the gulf countries of Bahrain. Regional culture is magnified by special design show, demonstrating the designer's fine. Pearl walls and long tunnel entrance, leading to an unknown fantasy world, everything is like a prelude to Alice in Wonderland.

The pursuit of surprise results Wanders, want to shop brings in a rich and full three-dimensional feeling, creating a constant changing scenery. After entering the shop, high-ceilinged space, minimalist use of color mixed with a strong Middle East style, giving a great visual impact. The main colors black and white gives a dignified, with a complex pattern of large area, dotted a small amount of blue glass windows and yellow walls, create a sense of dynamic and static-Safe, minimalist and complex collisions. Interior design focus are white performance. Its dome-type structure, echoing the dome of the local Islamic architecture, but also increased the level of a sense of space. Wall reliefs big white flowers, the beauty of asymmetry, giving a slightly oppressive shock.

↑ 空间一角的大试衣境 / Space corner of the big clothing throughout the trial.

↑珍珠门入口远景 / Pearl vision of the door entrance.
←镂空雕纹的通道 / Pierced carved patterns of the channel.
↓珍珠门入口细部 / Pearl door entrance detail.

↑ 别致的树形灯远景 / Unique vision of the tree lights.
↓↓ 蜂窝状橱窗和造型各异的人形模特 / Honeycomb windows and various shapes of the humanoid model.

↑ 室内空间 / Interior space.
↓ 平面图 / Plan

099

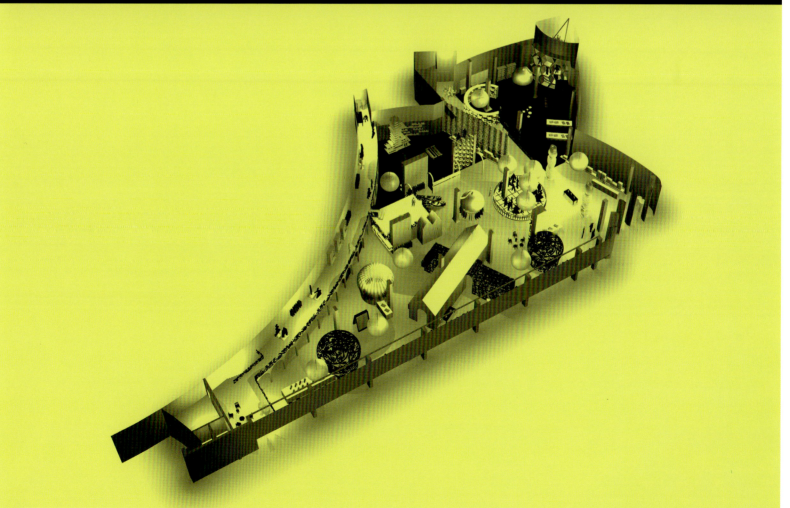

BRANCH
OF ROMANCE
杭州浪漫一身

16

【工程名称】杭州凤起路223号； 【面积】111 m²
【建筑设计】北京卅口建筑设计咨询有限公司； 【设计】迫庆一郎〔日〕
【主要用材】金属板网、石膏、白色地砖
【摄影】藤井浩司

露天剧院作为巴林的地标景观之一，以其颇具性格的混杂吸引着四方游客。Wanders决定将设计定位为"国际露天剧院"，即保留露天剧院混杂丰富的精神，注入现代的前卫奢侈元素，使之成为一个融古典中东风情和现代国际风尚于一体的设计空间。它更像是一个小城市，通过多种因素的综合，展现出建筑的美感。

Villa Moda入口处的墙面由众多巨大"珍珠"装饰而成，这些球形珍珠已经成了该店的标识之一。此外，长达10m多的入口通道的计灵感亦是来自巴林海湾国家的性质。地域文化通过别具一格的设计被放大展现，大气之中尽显设计师的细腻。珍珠墙和悠长的隧道入口，通向一个未知的奇幻世界，所有的一切都像爱丽丝漫游奇境的前奏。

追求惊奇效果的Wanders，希望店铺给人带来的一种丰富饱满的立体感，创造一种移步异景的应接不暇。进入店铺后，挑高的空间，极简的色彩运用混杂着强烈的中东风，给人极大的视觉冲击。黑白的主色调给人一种端庄克制之感，搭配以大面积复杂的花纹，点缀少量的蓝色玻璃橱窗和黄色墙面，制造一种动静皆宜的感觉，极简与繁复的碰撞。室内的设计重点均用白色表现。其穹顶式构造，呼应了当地伊斯兰的圆顶建筑，又增加了空间的层次感。墙面的白色浮雕大花，不对称的舒展之美，给人一种略带压迫感的震撼。

Open Theater as one of Bahrain's landmark landscape, with its rather mixed character of attraction for tourists. Wanders the decision will be designed as "an international open-air theater," that is a rich mixture to retain the spirit of open-air theater, infusing it with modern avant-garde luxury elements, making it a melting of classical style and modern Middle East, the international fashion in one of the design space. It is more like a small city, through a variety of elements, to show the beauty of architecture.

Villa Moda from the wall at the entrance to a huge number of "pearl" decorations made of these spherical pearls have become one of the store's logo. In addition, up to 10m of the total number of the entrance channel is also inspired by the nature of the gulf countries of Bahrain. Regional culture is magnified by special design show, demonstrating the designer's fine. Pearl walls and long tunnel entrance, leading to an unknown fantasy world, everything is like a prelude to Alice in Wonderland.

The pursuit of surprise results Wanders, want to shop brings in a rich and full three-dimensional feeling, creating a constant changing scenery. After entering the shop, high-ceilinged space, minimalist use of color mixed with a strong Middle East style, giving a great visual impact. The main colors black and white gives a dignified, with a complex pattern of large area, dotted a small amount of blue glass windows and yellow walls, create a sense of dynamic and static-Safe, minimalist and complex collisions. Interior design focus are white performance. Its dome-type structure, echoing the dome of the local Islamic architecture, but also increased the level of a sense of space. Wall reliefs big white flowers, the beauty of asymmetry, giving a slightly oppressive shock.

↑ 外立面 / Facade

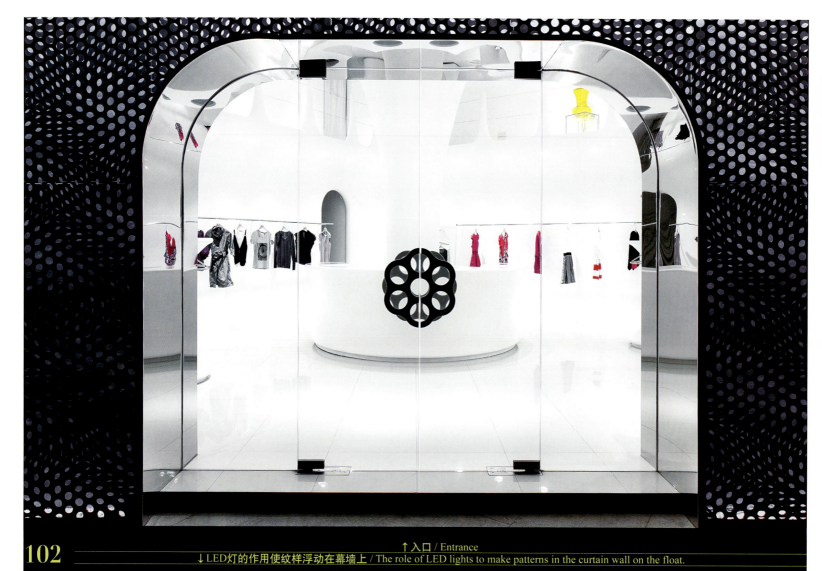

↑ 入口 / Entrance

↓ LED灯的作用使纹样浮动在幕墙上 / The role of LED lights to make patterns in the curtain wall on the float.

↑ 室内空间 / Interior space.
↓ 平面图 / Plan

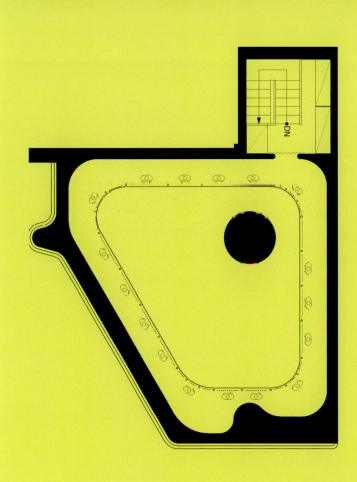
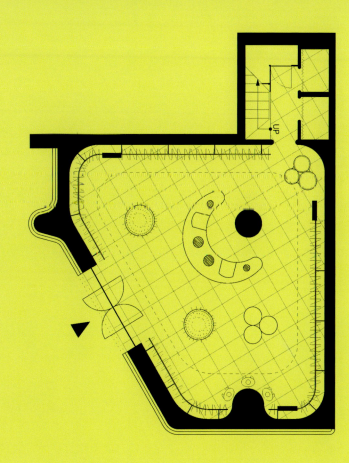

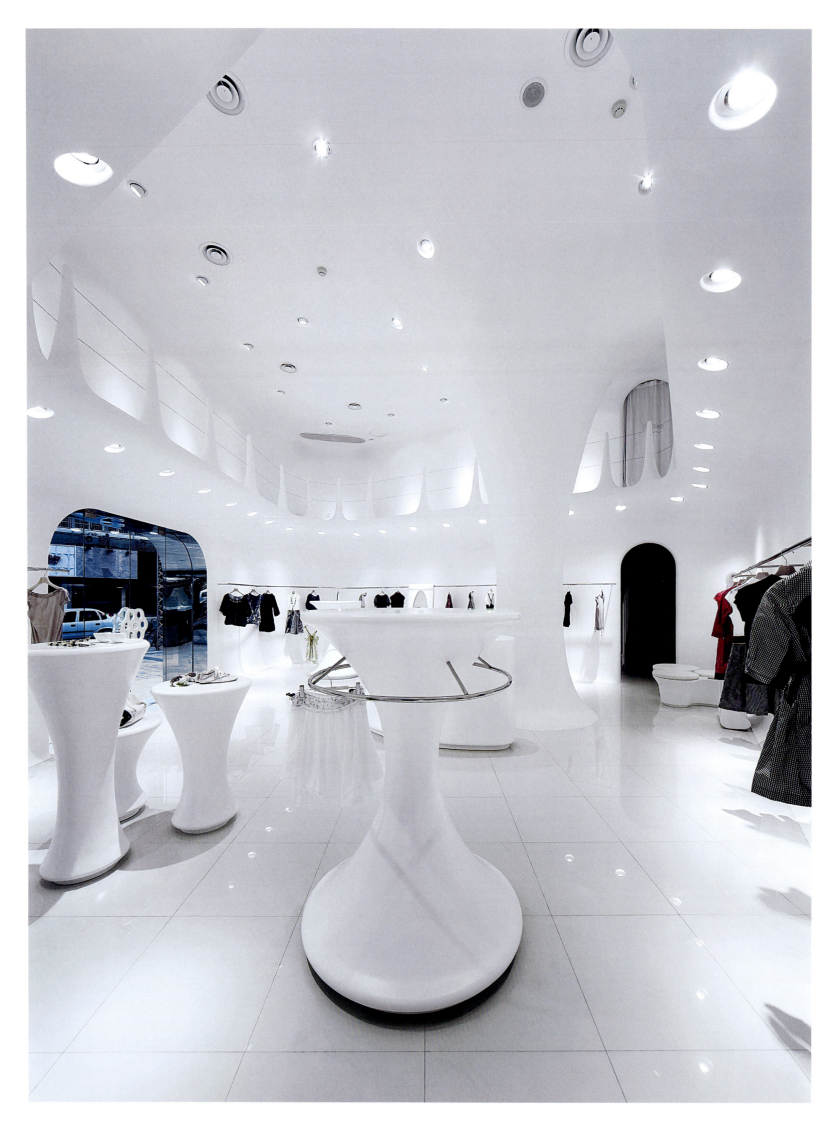

↑ 纯白的空间底色 / White space background.
← 超高的层高 / Space height.
↓ 剖面图 / Section

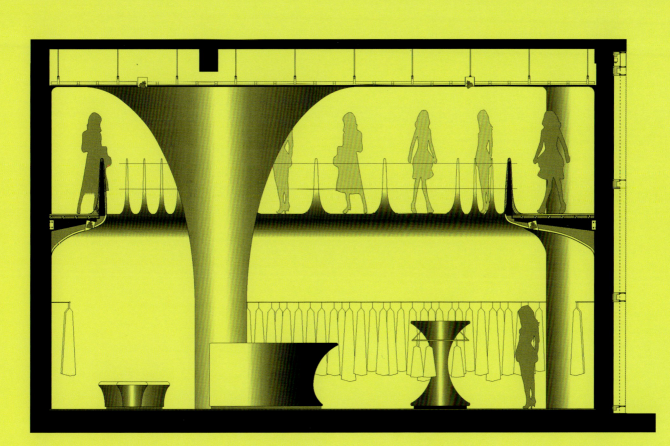

SPACE DESIGN IN GIANT SHOP
捷安特女性旗舰店

17

【坐落地点】中国台湾台北敦化北路；【面积】约280 m²
【设计】陆希杰；【设计公司】CJ Studio
【主要用材】毛丝面不锈钢、雾面石英砖、马赛克、人造皮
【摄影】Marc Gerritsen

设计师选用了柔美的曲线来表现这个充满女性特质的购物空间。整个店铺分为上下两层：地上一层面积较小，被合理的划分为教车区、展示区；地下室面积比较大，被分作4个区块的卖车区，即维修区、展示系统等。"地景拼图"是这个设计的主要概念。所谓"地景"即是利用各种空间陈设，设计出犹如自然地貌的模板，以此作为空间的分隔。而"拼图"则代表着各个休闲场景、活动的心情转换与联结，将此概念融入空间，表现于家具组合及天花。

由于此旗舰店主要是为女性打造的购物空间，所以在许多细节都有特别设计。一楼部分，退缩的门面让入口处有如舞台的框景，使阳光引入室内搭配室内轨道灯光，营造明亮的视觉空间。地下室为多功能空间，以活动组合家具串联起来，不但像是装置艺术，也同时创造停留地点，有休憩功能，使得空间增加了多功能性也更有趣味。

就空间色彩而言，设计师大量运用了黑、白、灰三种色彩，这样的颜色能够更好的衬托出产品的多样化和斑斓的色彩。主要用材方面，设计师使用了毛丝面不锈钢、雾面石英砖、马赛克、人造皮等材质，多种用材的选择让空间在保持整体性的同时在一定程度上延续了多样性的特质。

Designers selected to express the soft curves of femininity that is full of shopping space. The whole shop is divided into upper and lower levels: the ground floor area is smaller, has been classified as a reasonable driving area, display area; basement area is relatively large, was divided into four blocks of car dealers area, namely, maintenance area, display systems. "Landscape mosaic" is the main concept of this design. The so-called "landscape" is the use of space furnishings, design a template like the natural landscape as a space-separated. The "puzzle" is represented in various recreational scenes, activities, feelings of conversion and coupling, this concept into the space, furniture, composition and performance in the ceiling.

If the owner because of this flagship building of the shopping space for women, so many details are specially designed. On the first floor part of the back of the facade to the entrance to the stage like a window, so that the sun introduction of the indoor track with indoor lighting, to create a bright visual space.Basement multi-purpose space to the activities of combinations of furniture together, and not only like installation art, but also to create places to stay, there is open space features, making room for increased versatility and more fun.

The space in terms of color, designers make extensive use of black, white and gray color, this color to better bring out the diversification of products and gorgeous color. The main timber, the designers use the wool yarn surface of stainless steel, quartz tiles, mosaics, imitation leather and other materials, a variety of timber options for space in the same time maintaining the integrity of a certain extent, the characteristics of continuity of diversity.

↑ 设计师设计了很多展示台 / Designers a lot of Showcase.

↑活动组合家具串联空间 / Activity space for modular furniture series; ↑镜面的使用 / Mirror use.
↓天花以流线、自然有机的造型 / Ceiling in order to stream line, the natural organic shapes.

↑店面的主色调采用黑、白、灰色 / Store's main colors with black, white, gray.
↓平面图 / Plan

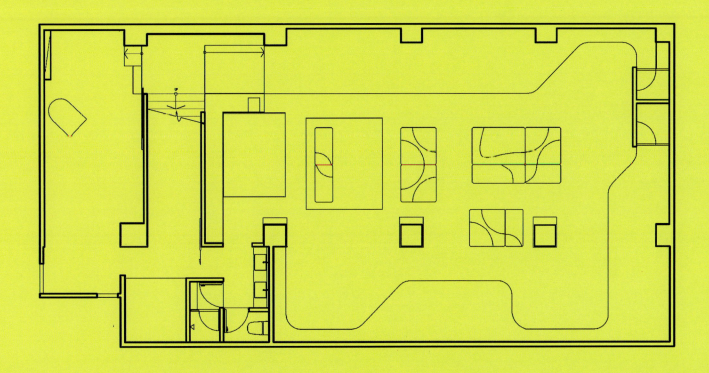

THE PIPE SHOW

Kymyka鞋包店

18

【坐落地点】荷兰马斯特里赫特；【面积】77 m²
【设计】Maurice Mentjens Design 〔荷〕
【合作设计】Angela Sanz, Agnisezka Cohojnacka, Raquel Rojas, Marta Melendo
【摄影】Courtesy Teresa Saey - A.L. Baltanás y Pablo Orcajo

整个商店空间是在两幢住宅的底楼，两幢住宅之间的墙被打通，楼层平面为镜像平面。Mentjens将建筑的垂直线条作为结构设计的基础。用柱子取代原来位于楼层平面交叉点的墙体。从地面到天花板的支撑钢梁均被镜子覆盖。

整个空间都被低矮的多功能展示台环绕包围。只有销售柜台像一座半岛一般，在整个空间显得较为突出。另外一组横梁有序地架构于木制的展示台上方，形成最简约的墙体框架，共计三层的横梁又成为向新巴洛克灰泥装饰的天花板的一种过渡。这些横梁呈线形排列，完全依照墙体结构的不规则形态架构，包围着整个空间。临窗的部分，横梁并没有隔断，形成临窗而立的小型展示架。

所有的室内元素（包括木制的展示台在内）均是定做的。在规划设计阶段，设计师对人体工学细部进行了充分的考虑，包括隔板门的开门方向、试鞋柜舒适的高度、甚至暗线路的可及性。设计师将展示台的底座稍稍后置，使得展示台看起来更加修长，契合商品品牌的内涵。每个枝形吊灯由9个卤素灯构成，这些吊灯被细钢丝悬挂在古老的中央天花板上，灯可以向任何方向旋转、倾斜，因此可以照亮所有的墙、展示台和钢管，使得每个产品能够达到被期望的展示效果。灯光经过反射产生光晕，为整个空间增添了些许神秘感。

The entire store space is in the two cities, on the ground floor, two homes have been opened up between the walls, floor plane of the mirror plane. Mentjens the vertical line of construction as structural design. To replace the original with the columns in the floor plane crossing the wall. The support from the ground-to-ceiling mirrors were covered by steel beams.

The whole space are surrounded by low-rise multi-purpose surround showcase. Only the sales counter, like a peninsula, the entire space look more outstanding. Another group of wooden beams and orderly framework for the showcase on the top of the wall formed the most simple framework, beams became the new Baroque plaster ceiling decorated with a transition. The beam was linear array, in full accordance with the structure of the wall structure of irregular shape, surrounded by the entire space. Window near the part of the beam is not cut off to form an independent mini-display.

All interior elements are custom-made. In the planning and design stage, the designer of ergonomic detail was fully considered, including the partition door open direction and try to shoe a comfortable height, and even dark line accessibility. Designers showcase the base of the post slightly to make showcase look more slender, fit brand connotations. Each chandelier from 9 halogen composition, these chandeliers are thin wire hanging in the old central ceiling. Light can rotate in any direction, tilt, so you can illuminate all of the walls, showcase and steel pipe, so that each product can be expected to reach the display. Halo lighting produced through the reflection for the entire space to add a bit of mystery.

↑窗户上的图案与店内陈设格调协调 / Windows on the coordination pattern and style furnishings store.

↑↑ 错落有致的鞋子散发出优雅的气质 / Patchwork shoes exude elegance.
↓ 中心展示区由两部分钢管展示区组成 / Center display area consists of two parts display area composed of steel.

↑ 低矮的多功能展示台环绕包围整个空间 / Low multi-purpose surround Showcase surrounded the entire space.
↓ 平面图 / Plan

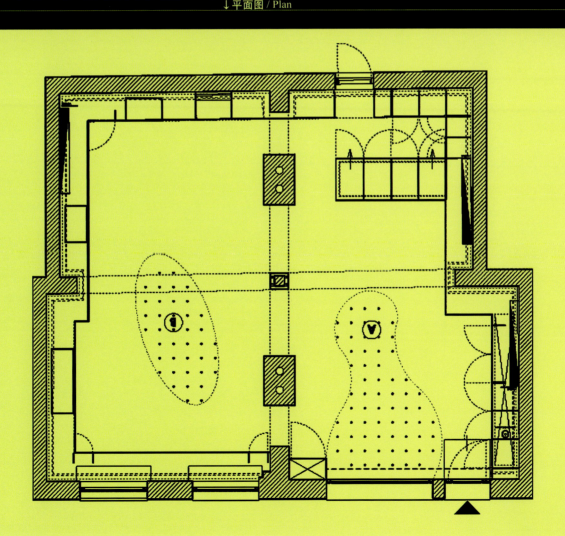

PAMPER
HEIRESS

Pamper Heiress旗舰店

19

【坐落地点】中国台湾台北市安和路；【面积】132 m²
【设计】陆希杰；【参与设计】曹车政、刘致佑、郑家皓
【主要用材】盘多魔地板、水泥漆、马来漆、镜面不锈钢、超白强化清玻璃、泰丝
【摄影】Maro Gerritoen

为了迎合该品牌的特点，设计师浪漫和精致定为设计的主旨，着眼于细节，将"Diamond House"的主题由内而外体现得淋漓尽致。"钻石"是贯穿整个案子的统一语汇和意象，中心以钻石造型为概念的珠宝展示柜是全场的视觉焦点，设计师以玻璃的切割拼接呈现钻石的多角反射，将展示柜中美轮美奂的珠宝映衬得愈加闪闪发亮。四周的展示柜也被切割为不同造型钻石，大大小小的"钻石"散落在空间每个角落，梦幻与现实之间似乎只有一线之隔。

为了与角状的珠宝柜台搭配，天花以折叠的线条与弧形的面做处理，切面与拱形在天花上交错，呈现出多种几何造型。黑色的木马在这个白色的空间并不显得突兀，它与空间在色彩上的对立强调了它存在的意义，激起对童年无忧无虑的快乐时光的回忆，设计师大量使用了水泥，并将这种原本粗糙冰冷的材料变得细致而亲切。在室外，灰色的水泥台阶沉闷而理性，而旁边一盏钻石造型的路灯，闪耀着璀璨的光芒，柔化并温暖了周遭的环境氛围；在室内，经过特殊的处理，地面、墙面已经无法看出水泥的原样，与空间的整体气质达到默契的和谐。

In order to meet the characteristics of the brand, designers will be romantic and refined as the design theme, focusing on the details, the "Diamond House" theme embodied most vividly from the inside out. "Diamond" is a case of the uniform throughout the vocabulary and imagery, the center diamond shape to the concept of jewelry display cases are the visual focus of audience, designer glass cutting and stitching show diamond reflection, a wonderful jewelry display cases in against the background were shiny. Around the display cases have also been cut into different shapes of diamonds, large and small "diamond" scattered in every corner of the space, between dream and reality it appears that only a thin line separated.

In order to match with the angular jewelry counters, ceiling to collapse of the lines and curved surfaces make handling aspect in the ceiling with arched cross, showing a variety of geometric shapes. Trojan black in this white space does not seem unexpected, it's opposite color space emphasized the significance of its existence, provoking a carefree childhood memories of happier times, designers extensive use of cement, and such original rough icy material that can become delicate and intimate. In the outdoors, dull gray concrete stairs and rational, but next to a diamond-shaped lights, shines with bright light, soft and warm atmosphere of the surrounding environment; in room, after special processing, the ground, the wall is no longer see the cement as it was, the overall temperament and space to reach a tacit agreement harmony.

↑ 从入口看向服装展示区和珠宝展示柜 / View from the entrance to the exhibition area clothing and jewelry display cases.

↑ 天花以折叠呈现多种几何造型 / Ceiling to collapse the lines showed a variety of geometric shapes.
↓ 天花的射灯与玻璃切面相映成辉 / Ceiling spotlights and glass aspect of interlaced.

↑ 钻石的意象贯穿了整个空间 / Diamond imagery runs through the entire space.
↓ 平面图 / Plan

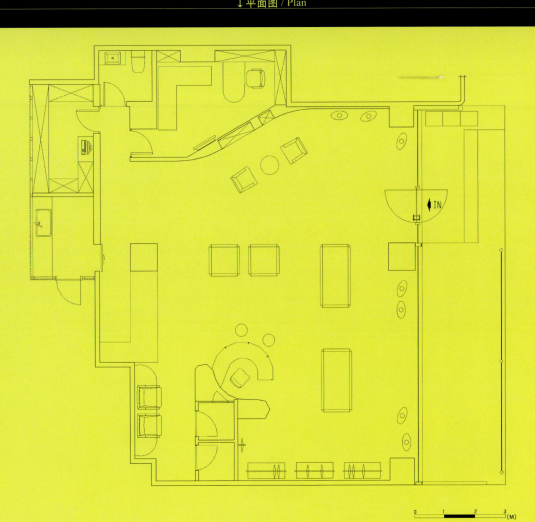

↑ 曲面的墙壁与天花衔接，空间中有了流动的感觉 / Surface of the walls and ceiling convergence space with the flow of feeling.
← 珠宝展示柜 / Jewelry display cases.
↓ 试衣间方便客人从各个角度欣赏试穿的效果 / Fitting room to facilitate the guests from all angles, try to appreciate the effect.

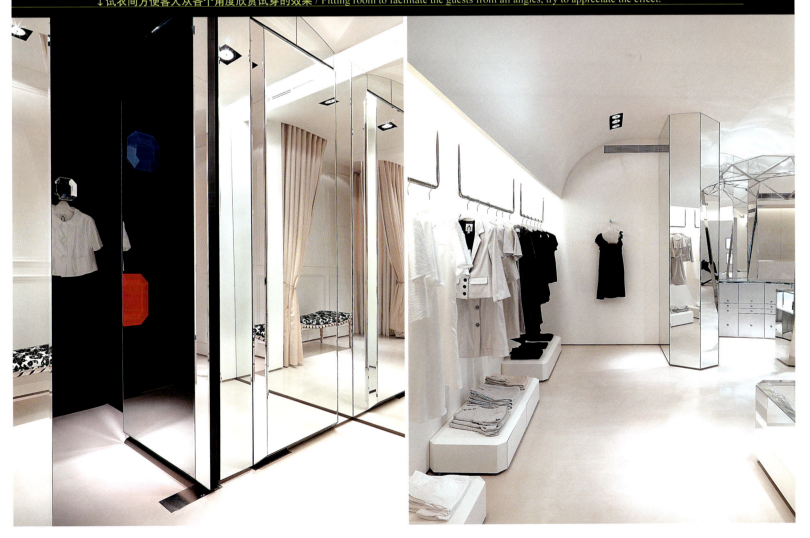

BACI
ABBRACCI
Baci & Abbracci 旗舰店

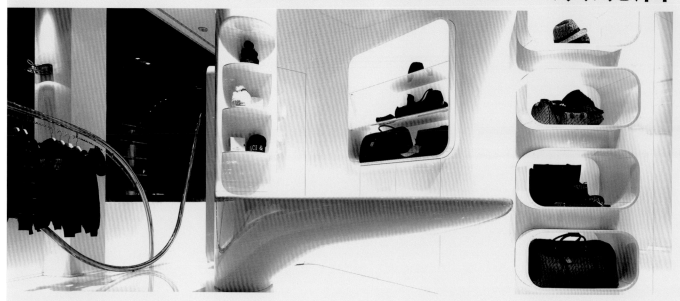

20

【工程名称】Baci & Abbracci 旗舰店
【坐落地点】意大利米兰
【设 计】Pierandrei Associated ［意］
【摄 影】Enrico Cano

这个设计里最有特色的两个部分是立面和橱窗的流线形态和空间内统一的白色调。白色的墙面采用了光滑发亮的材质，而形态上的蜿蜒曲折又使整个空间有种梦幻的感觉。正对入口的是半高的收银台，上方的挑高有2.5m，和起伏变化的立面形成了一个特有的三维建筑空间。而在临街的两排窗户旁边，设计师摆放了两个用不锈钢做成的闪亮的衣架。它们在空间的两侧，它们在空间的两侧，增加视觉上的横向尺度感和地台部分的感觉延伸。

3个试衣间像是在窗前伫立的3个发光的圆柱子：光线穿过垂下的棉絮和小块的透明树脂玻璃，不锈钢管把试衣空间分隔出来。设计师在这里做了一个视觉上的游戏——"看见或看不见"。顾客将在这个若隐若现的半透明空间里感受试衣带来的愉悦过程，好好体会这一个动人的时刻。为了增强触觉和视觉的效果，设计师采用了高光的复合材料。白色和镜面般光亮的不锈钢管将室内的灯光反射，增强了展示的效果。室内惟一有色彩的是结合平面形成的飞梭形的天花。在透光板后面有一组LED灯，由电脑系统控制的RGB色彩系统能使天花发出各种不同的颜色。设计师给这套装置起了个名字——"plug-ins"，纯色的照明能适应平时商店的购物气氛，变化的彩色天花到晚上又能和周围的餐厅、俱乐部争奇斗艳。

This design is the most distinctive facade and windows, a streamlined and unified state white tone. White walls with a smooth shiny material, while the morphological meandering again so that the entire space a kind of dream feeling. Being imported is the half-height of the cashier, the top of the high-ceilinged with 2.5m, and the facade form a unique 3D architectural space. In the street next to the two rows of windows, designers placed two made of shiny stainless steel hangers. They are in space on both sides to increase the visual sense of the horizontal scale part of the feeling of an extension of the platform.

Fitting room like standing in the window of the cylindrical light-emitting sub-3: hanging light passes through the cotton, and small pieces of transparent resin, glass, stainless steel tube to separate out the dressing room. Designers are here to do a visual of the game - "see or not see." Customers will be looming in the semi-transparent space feel pleasure to bring the process of dressing, well understand this a touching moment. In order to enhance the tactile and visual effects, designers used a high-light composite materials. White and mirror-like polished stainless steel indoor lighting reflection, enhances the display effect. The only interior color is the combination of a flat form of smallpox. Translucent panels in the back of a group of LED lamps, controlled by the computer system RGB color system allows the ceiling to issue a variety of colors. Designer to the device, gave a name - "plug-ins", pure color lighting to adapt to normal store shopping atmosphere, changing the color of smallpox into the night but also in and around restaurants, clubs and contests.

↑ 建筑外观 / Architectural appearance.

↑ 衣架的曲线设计 / Racks curve design.
↓ 室内展示柜 / Indoor display cabinets.

↑ 3个试衣间相连 / 3 dressing rooms attached; ↑ 墙面从上到下都用来展示 / Walls from top to bottom are used to display.
↓ 平面图 / Plan

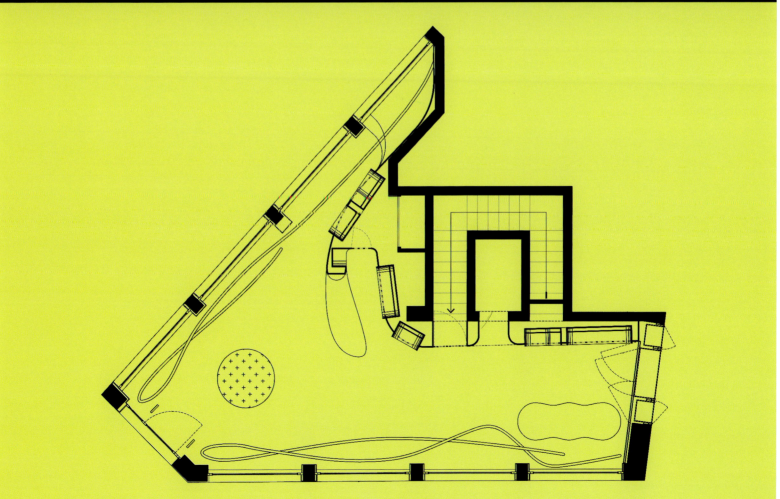

COMBINATION
OF SIMPLICITY AND FASHION
Fancl ifc全新概念店

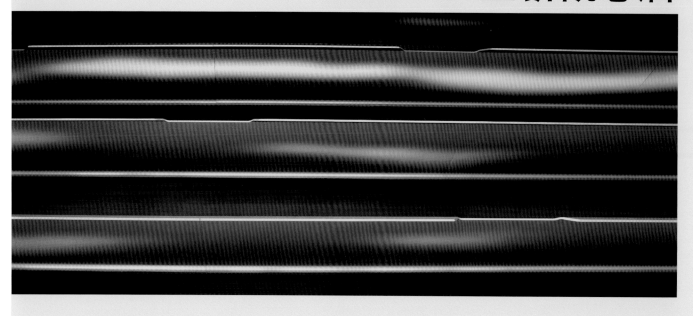

21

【项目名称】FANCL ifc全新概念店
【坐落地点】香港中环国际金融中心商场1086号铺
【面积】392 m²
【设计】Gwenael Nicolas 陈幼坚

全店以极富格调的FANCLBlue及银色为主调，清新而时尚。此外，设计师非常重视货品陈列架的设计和质量，为突出简约线条，特地从日本订造高科技金属层板，外形与内涵并重。由于选用的物料、色温以及灯光效果均会对FANCL Blue的色调及效果有所影响，因此我们利用尖端科技，经过逾千次的测试，以确保店内所有的蓝色均一并带出美感。另一挑战就是制造高科技金属陈列架，为达到FANCL超高的标准，所有制作包括物料、模型等均于日本完成并空运抵港，仅厚度和弧度已经过多次改良，务求令每件陈列架的水平差距近乎零。

简约的线条正是店铺的焦点之一。踏进店内，货品井然有序地整齐排列在一行行的银色高科技金属层架上，与置放在店内中央的产品展示及皮肤测试分析区相互呼应。为了突显无添加的特质，设计师刻意摒弃店内多余的摆设，简约内敛，充分体现产品的特色陈列方面，先进的设计更把产品巧妙地隐藏在回卷式半透明蓝色小玻璃门后，设计功能性与观赏性兼备。此外，实践"以客为先"的优质服务承诺，概念店内还设有"个人专业美容咨询区"，让顾客可以安坐于日本订制的时尚银灰软椅上，享受个人化的专业美容咨询及护理。

All shops very tone of FANCLBlue and silver-based adjustment, fresh and stylish. In addition, the designer goods display rack attaches great importance to the design and quality, in order to highlight the simple lines, specially from the Japanese custom-made high-tech metal laminates, appearance and connotation of equal importance. Result of the selection of materials, color temperature, and lighting effects are FANCL Blue will have an impact on the tone and effect, so we use cutting-edge technologies, through thousands of tests to ensure that all of the blue line store. Another challenge is to create high-tech metal display racks, in order to achieve the FANCL high standards, all of the production, including materials, models, etc. are completed and air arriving in Japan, only the thickness and curvature have been modified several times, each display shelf the gap between the level of almost zero.

Simple lines is the focus of one of the shops. Enter the store, goods were neatly arranged in an orderly manner in line after line of high-tech metal layer of silver racks, and placed in the central store products and skin test analysis zones each other well. In order to highlight the characteristics, designers intentionally discard the extra furnishings store, simple restrained, fully embodies the characteristics of the product display area, the product of advanced design more cleverly hidden in the back to roll translucent blue glass door, designed to function and both spectator. In addition, the practice of "Customer First" commitment to quality service, the concept store has also set up "personal and professional beauty consultation with the District," so that customers can sit in the Japanese custom of fashion silver gray soft chair and enjoy the personalized professional beauty consultation and care.

↑ 概念店犹如太空舱一般 / The general concept store like capsule.

↑ 货架横向透视 / Shelf horizontal perspective.
← 光线在空间内的变化 / Light in the space changes.
↓ 平面图 / Plan

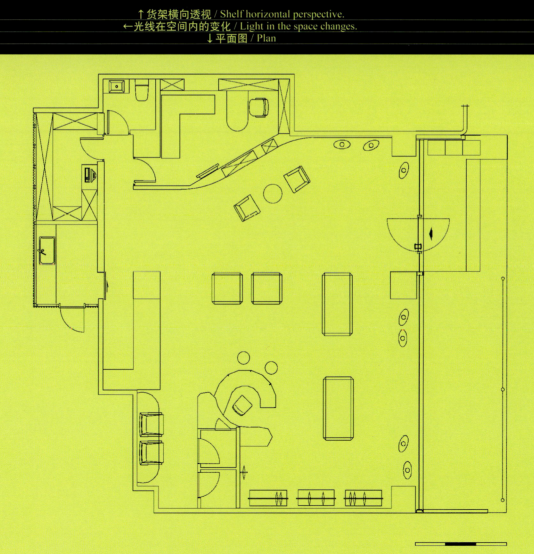

↑↓ 货架细部 / Shelf detail.
→直线贯穿的走向，让Fancl全新的概念店显得格外深邃 /
A straight line through the direction, so that the new concept store Fancl extraordinarily mysterious.it is structured.

MAISON MARTIN MARGIELA
MMM专卖店

22

【坐落地点】意大利米兰 Via della Spiga for MMM
【面积】180 m²
【设计】Martin Margiela 〔比〕
【摄影】Courtesy of Martin Margiela

白色是Maison Martin Margiela的灵魂，木质材料和白颜色总是出现Maison Martin Margiela的各个角落。可别把这里的白色和医生、科研人员身上的白色相提并论，那些都太严肃。对于Maison Martin Margiela来说，白色代表了战胜脆弱的力量，它像是一块白布，让设计师们毫无约束地大胆表达自我。而且白色确实是包容性最好的颜色，可以和任何颜色搭配，而"混合"本来就是Maison Martin Margiela的理想。

整个店铺约为180 m²，包括男女服装、配饰和一些非卖品。如果说，空间的设计追寻了某种独特的哲学思想或者文化的话，店铺里的各个角落都反映了对不同文化的膜拜。在"混合"的主张下，很多不相干的东西都摆在了一起。当然，在做空间设计的时候，还是要好好利用透视法则。店铺里面有许多旧家具，充当货架和展示品，还有一些从各地收集的各色玩意。在其中的一些空间里，旧窗户被拼接在一起，作为天花，但是没有玻璃，非常独特。而在前厅，许许多多来自香港的旧书籍被刷成白色，堆砌起来形成隔墙，可以完全独立的展示货品。

White is the soul of Maison Martin Margiela, wood materials, and white colors always appear in every corner of the Maison Martin Margiela. Do not white could be here, and doctors, researchers compared the white body, those who are too serious. For the Maison Martin Margiela, the white represents the fragility of the power to overcome it like a piece of white cloth, so that designers without any restraint, bold self-expression. And the white is really inclusive, the best color, can match any color, while the "mixed" Maison Martin Margiela has always been ideal.

The whole shop is about 180 m², including male and female clothing, accessories and some non-sale. If we say that the design of space to pursue a kind of unique culture, then every corner of the shop reflects the worship of different cultures. In the "mixed" claim, many irrelevant things are put together. Of course, when doing space design, or want to make good use of perspective rules. There are many shops inside the old furniture, serve as shelves and display products, there are collected from all over the colored stuff. In which some of the space, the old windows were spliced together, as smallpox, but no glass, is unique. In the lobby, many of the old books from Hong Kong have been painted white, piled up to form walls, can be completely independent of the display of goods.

↑整个隔墙都是以书堆砌而成 / Are based on the entire wall built on the book.

↑ 简单的橱窗 / A simple window.
↓↓ 整齐的男装区 / Men's neat area.

↑ 室内空间 / Interior space.
↓ 横竖之间充满了线条感 / Between the lines if they had full feeling.

POPLAR KID'S REPUBLIC
上海蒲蒲兰绘本馆

23

【坐落地点】上海静安区愚园路262号；【面积】一层37 m², 二层68 m²
【设计】迫庆一郎〔日〕
【参与设计】Shuhei AOYAMA、Yu FUJITA、SAKO Architects
【摄影】沈思海

这是一家绘本馆，一楼是用作书店使用，二楼作为活动空间和办公室。一棵大树，将其解体成各个要素，并将这些要素嵌置于空间之中。树干、树枝、树皮和树叶，对其进行分割、加工从而抽象化的同时，并不失去其本身的物质感。于是这些素材变成天花、墙壁、楼梯和家具等空间要素被重新构建起来。有点像一条鱼切分成几块，每个部分烹饪后又重新组合起来的过程。

一楼，书柜排满了一整面墙。使用各种树木的板材做出宽度、深度、厚度各异的书柜，它们构成了一个立体的构造。其中又有用挖空的一段原木做成的书柜，由此带来新的"表情"。地板上投影出繁茂的树叶，让人感受到透过枝叶倾洒出的阳光。天花上等间距排列着带有枝杈的原木，从街道上走过，透过玻璃便可看见树木一端年轮的模样。

二楼，便可看到由粗粝的树皮覆盖的半球形的活动空间。内部是纯白的好像雪屋一样的空间，顶上开有几个不规则圆洞。地面做出高差，作为舞台及观众席来使用。铺上彩色地毯，孩子们可以自在地躺在上面阅读。穿过活动空间向里便是办公区域。办公桌上铺有切成片的原木截面，并一直延伸到墙面上。卫生间的风格完全不同，各面都使用镜面不锈钢作装饰。走进其中的人，在镜面不锈钢的反射下被无限的复制开来。

This is a painting museum, on the first floor is used bookstore to use, on the second floor as a space and offices. A tree, breaking into its various elements, these elements will be placed in the room inlay. Tree trunks, branches, bark and leaves, its segmentation, processing, and thus abstract the same time, do not lose a sense of their own material. A result, these materials into ceilings, walls, stairs and furniture, etc. were re-built up space elements. A bit like a fish, cut into pieces, each part of the re-combined after cooking process.

The first floor and a bookcase filled the entire wall. Using a variety of trees to make the width of the plate, depth, thickness varying bookcase, they constitute a three-dimensional structure. And among them, made by hollowing out a section of wood of the bookcase, and the resulting new "expression." Projection on the floor out of lush leaves, people feel the sun spills through the branches. Upper ceiling lined with a twig spacing logs, from the streets through it, will see the trees through the glass at one end-ring shape.

On the second floor, can see from the rough bark-covered domed space. The interior is pure white like snow, like house space, there are several irregular hole on top to open. Make a height difference on the ground, as the stage and auditorium to use it. Covered with colored carpet, the children can freely lie to read. Through to the inside space is office space. Desk upper berth logs are sliced cross-section, and has been extended to the wall. The bathroom's style is totally different from the surface of stainless steel mirrors are used for decoration. Walk into one of the people at the mirror reflection of stainless steel has been opened to unlimited copying.

↑木框架粘在表盘一样的圆盘上 / Wooden frame glued to the dial on the same disc.

↑ 通向二层的楼梯 / Staircase leading to the second floor.
↓ 一层平面图 / 1st floor plan.

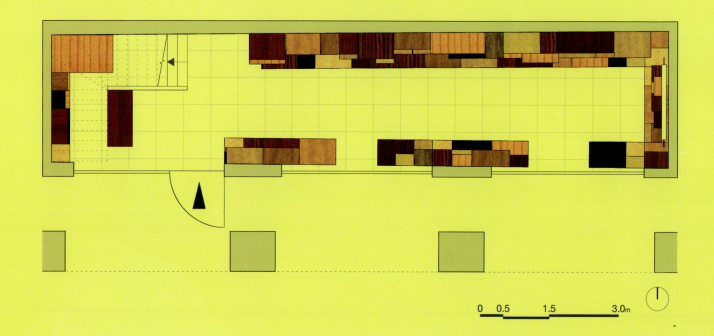

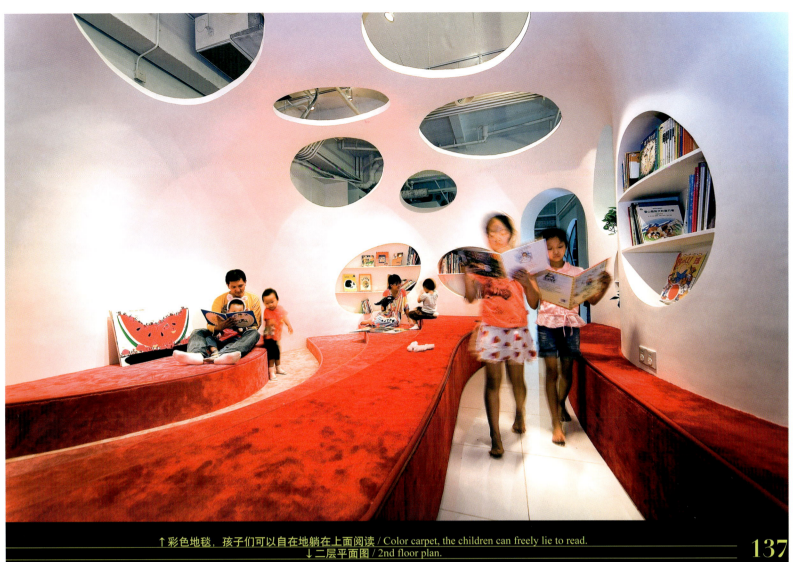

↑ 彩色地毯，孩子们可以自在地躺在上面阅读 / Color carpet, the children can freely lie to read.
↓ 二层平面图 / 2nd floor plan.

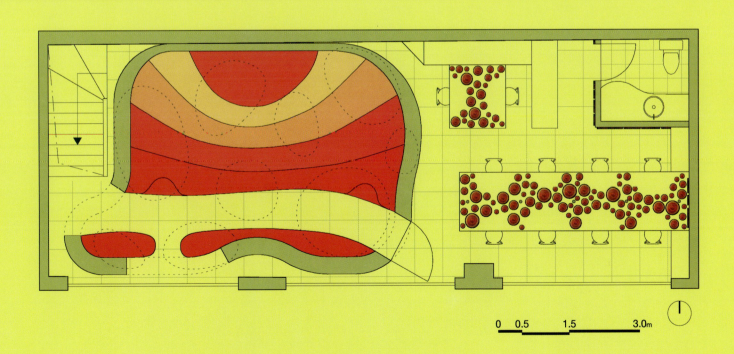

↑ 各种形状的木质书架平添了很多童趣 / Wooden shelves of various shapes adding to a lot of playful element.
↓ 可以看到用粗粝的树皮覆盖的半球形活动空间 / Can be seen covered with thick coarse bark of the hemispherical space.

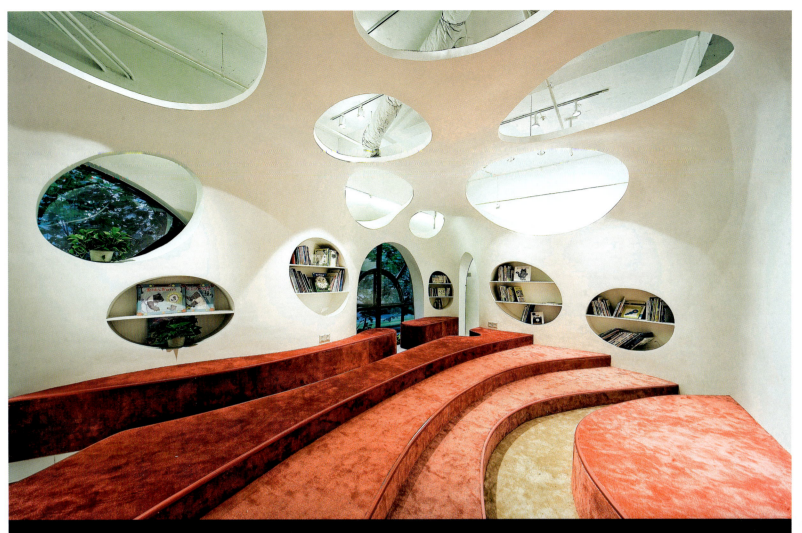
↑ 室内空间 / Interior space.

↓ 卫生间各面都使用镜面不锈钢作装饰 / Bathroom mirror stainless steel on each side are used for decoration.

MARNI
FLAGSHIP

Marni品牌旗舰店

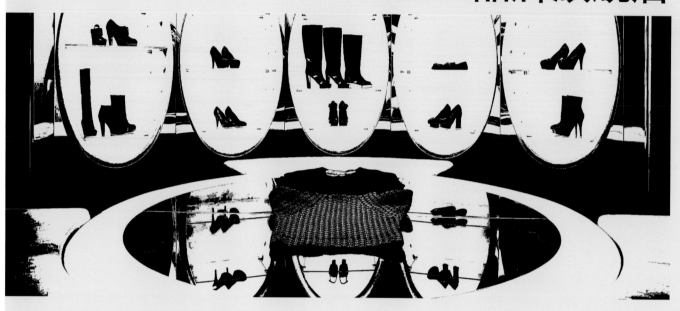

24　【坐落地点】美国纽约上东区麦迪逊大街159号；【面积】128 m²
　　　【设计】Sybarite建筑事务所〔英〕
　　　【主设计师】Simon Mitchell, Torquil McIntosh
　　　【参与设计】Massimilliano Tiezzi, Susan Heffernan, Doukee Wang

旗舰店是展示品牌形象的重要形式，因此无论商家还是设计师，都为此投入了比设立一般零售店更多的精力。店铺原本是一个十分狭长的空间，设计师特意将隔壁空置的房间也纳入到设计中来，赋予这两个空间不同的布局和功能，并且用一种"树形不锈钢管"贯穿全局。

抛光不锈钢的大面积使用为空间注入了非凡的效果。巨大的椭圆形的抛光不锈钢吊顶像一朵云彩"飘浮"在上空，将室内空间高度和纵深进一步拉大，无形中空间被放大数倍。吊顶内部设置数个射灯，以满足照明的需要。色彩艳丽的波西米亚造型元素由墙壁倒映在吊顶上，上下连成一体，顾客随时随地融入现代艺术的包围之中，时尚而浪漫的气息无处不在。新扩充进来的一半空间原本有一面十分厚实而笨重的墙体，在设计师巧思妙想之下变成了一面轻盈的产品展示墙，一半是采用玻璃纤维的透明墙，另一半墙体则以不锈钢衬底。空间的尽头是一个圆弧形的鞋子展示架，仍旧是抛光不锈钢与玻璃纤维的经典组合。天花上4个巨大的圆形的镂空，配以法国高品质透光软膜照明设备。若干条细如发丝的白线顺势垂下，上面挂着的服装好像"飘浮"在半空中，雅致而飘逸，将服饰的美感充分体现出来。

Flagship store is to demonstrate an important form of brand image, so they businessmen or the designers have put in a general retail store than to set up more energy. Shops used to be a very narrow space, designers specifically to the vacant room next door is also incorporated into the design to give these two spaces with a different layout and functionality, and using a "Tree of stainless steel pipe" through the global.

Polished stainless steel used for the large area of space into extraordinary results. A huge oval-shaped polished stainless steel ceiling, like a cloud "floating" in the sky would be the height and depth of interior space and further widening, virtually space is magnified several times. Ceiling spotlights a number of internal settings in order to meet the lighting needs. Colorful bohemian styling elements from the wall reflection in the ceiling, the upper and lower fused into the customer at any time surrounded by modern art, fashion and romantic atmosphere everywhere. The new expansion of the incoming half of the space originally had a very thick and heavy side of the wall, under the designer ingenuity would like to become a wonderful lightness of the products side walls, half of the transparent glass fiber wall, while the other half of the wall stainless steel substrate. Space is an arc-shaped end of the shoes display, polished stainless steel and glass is still the classic combination of fiber. Ceiling on the four large circular hollow, accompanied by French high-quality translucent soft-film lighting equipment, a number of hair-thin white line of homeopathic hanging above the clothing hanging like a "floating" in mid-air, elegant and elegant, the beauty of the dress is fully manifested.

↑ 不锈钢管在空中优美地回旋 / Stainless steel tube in the air gracefully maneuver.

↑↑ 不锈钢吊顶将空间拉伸 / Stainless steel ceiling space stretching.
↓ 服饰仿佛飘浮在半空中 / Dress as if floating in the air.

↑ 服饰仿佛飘浮在半空中 / Dress as if floating in the air.
↓ 平面图 / Plan

↑ 不规则的盒子排布在墙上 / Irregular row of cloth on the wall of the box.
← 艳丽的背景墙倒映在吊顶上 / Colorful backdrop reflection in the ceiling.
↓ 圆弧形的鞋子展示架 / Arc-shaped shoes display.

ONE2FREE
FLAGSHIP SHOP

One2free旗舰店

25

【坐落地点】中国香港；【面积】500 m²
【设计】陈幼坚
【设计团队】Alan Chan Design Co, Gwenael Nicolas
【摄影】Alvin Chan

一切自然的本源是"水"，设计者用他来展现整个卖场简洁、灵动、清澈的空间内涵，并传达出One2free"先进、活力、创新、可信、亲切"的品牌精神。"水"的概念被运用到商店的整体形象和展示陈设上，店堂中心两个上下连接的360°巨型水瓶装置由日本原创设计并制作，丙烯酸酯材料做成，是专门用来做产品展示、交互式示范和广告宣传之用的。

这个设计通透、清澈，在体现了"水"的特质的同时更方便了顾客浏览产品，高科技感应运而生，是整个室内设计的点睛之笔。透明柱瓶四周环绕的四扇橙色玻璃屏风，则像能自动开闭的太空舱门，使整个产品展示空间更显精致、神秘和未来感。而最富情趣的是从繁忙的一楼展厅一直向三楼耸立着的不锈钢环形指示装置，仿佛从水下涌出的气泡，向上轻轻漂浮着，通向二楼的楼梯旁是一片刻有水波纹的镜面墙体，它如一片宁静的湖面将卖场里的一切都倒映其上。"水"的意境充满了整个空间。

全新的客户服务专柜采用创新的双屏幕设计，一长条纯白色的光滑台面上镶嵌了电子屏幕，使顾客可以从个人的屏幕中一边观看最新的服务推广信息，一边处理账户事项或核对交易，简洁硬朗的体验空间被消费者所喜爱。

All the natural origins of "water", the designer to show the entire store with his simple, Smart, clear space for content, and convey One2free "advanced, dynamic, innovative, credible, friendly," the spirit of the brand. "Water" concept has been applied to the store's overall image and display furnishings, the center two upper and lower their stores connected to 360° giant bottle device and produced by Japan's original design, made of acrylic material, is designed to make products , interactive demonstrations and advertising of the use.

The design is transparent, clear, in the concept of "water" character at the same time more convenient for the customers to browse products, high-tech sense came into being, is the crowning touch to the interior design. Transparent column surrounded by four bottles of orange glass screen, automatically opening and closing like a door space, so that the entire product display space is even more refined sense of mystery and the future. The most fun was busy on the first floor from the hall had been erected to the third floor of the stainless steel ring pointing device, as if from underwater gush of bubbles floating gently upwards. Near the stairs leading to the second floor is a moment of water ripple mirror wall, it will be like a tranquil lake, everything inside the store reflected upside from using it. "Water" mood filled the entire space.

The new customer service counters innovative dual-screen design, a strip of pure white mosaic of the smoothness of the counter electronic screen, so that customers from the individual side of the screen watching the latest service promotion while dealing with matters or checking account transactions , simple experience of space is loved by consumers.

↑通向二楼的楼梯旁是一片刻有水波纹的镜面墙体 / The stairs leading to the second floor is a moment next to the mirror wall of water ripples.

↑ 一楼展厅 / On the first floor exhibition hall.
↓ 一层平面图 / 1st floor plan.

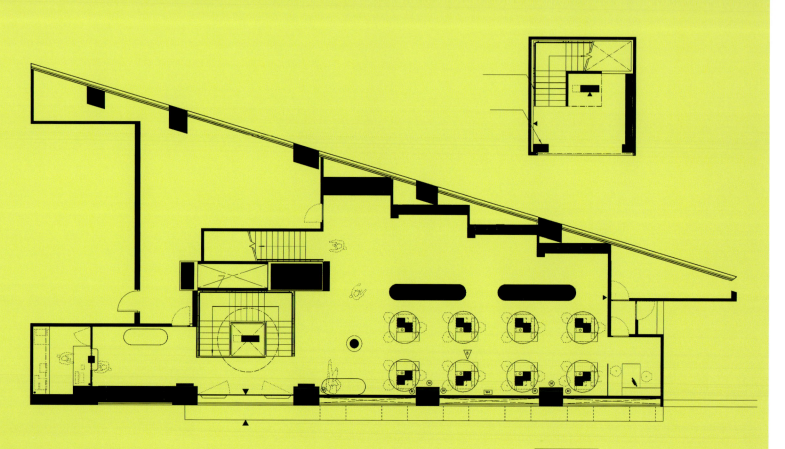

MF PLAN

↑ "水"的概念被运用到商店的整体形象设计上 / "Water" concept has been applied to the overall image of the store design.
↓ 二层平面图 / 2nd floor plan.

↑ 纯白色的光滑台面上镶嵌了双屏电子屏幕 / Pure white, smooth countertop mosaic of the dual-screen electronic screen.
↓ 黑与白的交融 / A blend of black and white.

↑ 产品展示空间 / Product exhibition space.
↓ 展示空间细部 / Exhibition space detail.

VILASOFA
FLAGSHIP STORE
Vilasofa旗舰店设计

26

【坐落地点】Barendrecht，荷兰；【面积】1400 m²
【设计】Tjep.〔荷〕
【参与设计】Frank Tjepkema, Janneke Hooymans, Leonie Janssen, Tina Stieger, Bertrand Gravier
【摄影】Courtesy of Martin Margiela

这原本用来展示商品的地方在设计师的巧手雕琢之下即刻充满了家庭般温馨的气氛。这里出现的每一个设计元素和摆设都来自VilaSofa自己，这些点滴汇聚起来便营造出一个独特的空间来。

Tjep.在对于设计概念的阐述简单的令人惊讶，但却非常合乎逻辑，那就是：将店铺设计和家居设计的美学观念结合在一起。因此，即便是最普通的材质，哪怕是胶合板，在这里都能得到很好的运用。另外，这里随处可见那些常运用在货品运输和包装上的标志，它们在这里被演绎成为一种装饰元素，不断地重复变换着出现，用以划分地界、安排路径和组织空间。商店的墙面设计也很特别，设计师在一面巨大的墙上镂空雕刻了一座漂亮的豪华别墅的剪影。相信不少客人看到它都不由自主地想到了自己的家。墙上那些曲线动人的树枝形装饰灯架、形态各异的缺口、多样有趣的窗户、极致浪漫的装饰，以及漂亮的类似装饰画等的剪影，都让人心动不已。

在这个空间里，Tjep.还设计了一组类似野餐用的大桌子供客户跟店员交流。当然也少不了温暖舒适的小包间，这里更加私密舒适。

This was originally used to show where the designer goods under the skilled carving instantly filled with a warm family-like atmosphere. Here every design elements and furnishings are from VilaSofa themselves When these clusters would be to create a unique space.

Tjep. In the design concept for the elaboration of the simple surprising, but very logical, that is: the shop design and aesthetics of home design ideas together. Even the most common material, even if it is plywood, where access to good use. In addition, there can be seen everywhere who often used in goods transport and packaging on the signs, where they have been interpreted as a decorative element, to repeat changing the appearance for the division of boundaries, arranging the path and the organization of space. Store very special wall design, designers in the side walls of a huge sculpture of a beautiful silhouette of the luxury villas. Believe that many of the guests to see it all automatically think of his own home. The branches are touching the wall curve-shaped decorative lighthouse, gaps of various shapes, various interesting windows, the ultimate romantic decoration, as well as a beautiful silhouette similar to the decorative painting, etc., are enticingly endless.

In this space, designers also designed a set of similar large picnic tables for customers with the exchange of staff. Of course, ultimately, warm and comfortable rooms of packets, where a more intimate comfort.

↑墙面伸出的红色小滑梯 / Slide out of the red wall.

↑ 主题墙面设计了一座镂空的豪华别墅的剪影 / Theme designed a hollow wall luxury villas silhouette.
↑↓↓ 温暖舒适的小包间 / Warm and comfortable rooms packet.

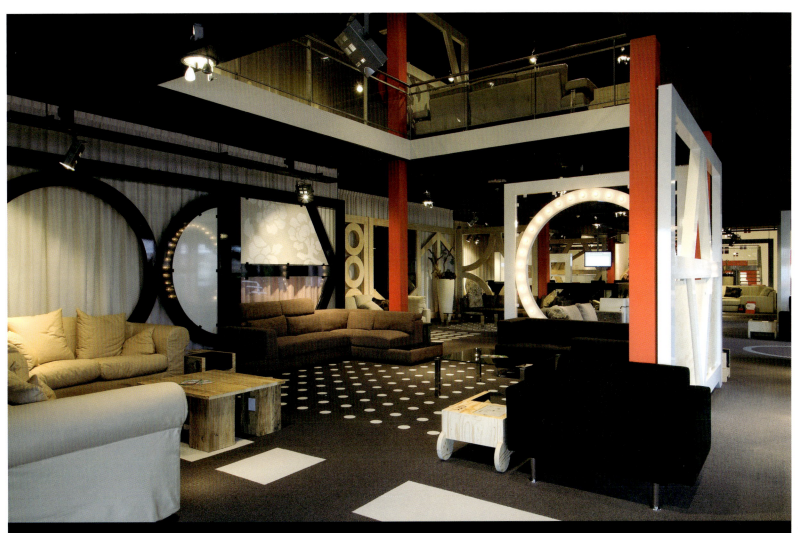

↑ 简单的设计元素被用来组织流线和安排空间 / Simple design elements be used to streamline the organization and arrangement of space.
↓ 平面图 / Plan

LINDEN APOTHEKE

Linden Apotheke药店

27

【坐落地点】德国路德维希堡
【设计】Ippolito fleitz group〔德〕
【材料】花岗岩鹅卵石、白色环氧树脂、地毯、白色漆、中密度纤维板
【摄影】Zooey Braun

设计师的设计风格带有典型的德国风范，这个空间设计严谨，线条简约而不简单，纯粹的白色墙面、连续精致的货架和墙面的弧线等这些元素组合在一起，赋予这里非凡的面貌。房间的整体性通过从墙壁到大花板光滑的弧形过渡及白色墙面的连续性得到进一步的强调。地面部分也是设计师最精心研磨的部分之一，花岗岩鹅卵石这种现代设计极少运用的材质被提炼出来，它反映的是典型的巴洛克风格。为了充分利用空间，药店中间部分还设计了三个可旋转的商品陈列柜，以便人们更方便的挑选商品。

药店的设计亮点在于店内艳丽的主题彩绘，它是由Ippolito fleitz group的设计师同纺织品设计师莫尼卡•特仑可拉（Monika Trenkler）共同设计，上面描绘的是11种药材的壁画，而这11种药材也恰恰就是药店的象征图案。设计师在这里通过简练的色彩语言对空间传统主题进行阐述：当你生病的时候，药草是必需的，这些药草通常和特殊的情境、故事相联系。因此，天花板主题里的11种药草出现在了人们的眼前，每一种药材都代表一个故事。

室内设计中弧形、壁画元素和花岗岩鹅卵石带来了些许怀旧的感觉。而它的设计手法却是现代的、简约的。设计元素的交替使用，体现了药店的主题，让来到这里的人感到亲切与愉悦的同时，也对品牌有了更加深刻的认识。

The designer's design style with a classic German style, this space is well-structured, simple lines rather than a simple, pure white walls, a continuous refinement of the shelves and walls of the arc and so these elements combine to give here an extraordinary face. The integrity of the room from the smooth curved walls to the ceiling and white walls of the continuity of the transition has been further emphasized. The ground segment is also the designer of the most elaborate part of the grinding of the granite cobblestones that very few modern design has been refined out of the use of the material, it reflects the typical Baroque style. In order to make full use of space, the middle part of the pharmacy also designed three rotating showcase of goods so that people more convenient selection of merchandise.

Pharmacy is designed to highlight the theme of the shop painted bright, it is by Ippolito fleitz group of designers with the textile designer Monika Trenkler joint design, the above depicts the 11 kinds of herbs murals, which are also 11 kinds of medicinal herbs is precisely the symbol of pharmacy pattern. Designer colors concise language adopted here the space of traditional themes: When you are sick, when herbs are needed, these herbs are often, and special situations, the story linked. Therefore, the ceiling inside the 11 kinds of herbs theme appeared in people's eyes, each of which represents a story of medicine.

Curved interior design, murals elements, and brought a little granite pebbles nostalgic feeling. And its design technique is modern, simple. The alternate use of design elements, reflecting the theme of the pharmacies, so that people who come here are friendly and pleasant at the same time, also have a more profound understanding of the brand.

↑入口处的弧线突出了空间的特点 / Arc at the entrance to highlight the space character.

↑ 药店整齐干净 / Pharmacies neatly trimmed.
→ 主题彩绘 / Theme painting.
↓ 平面图 / Plan

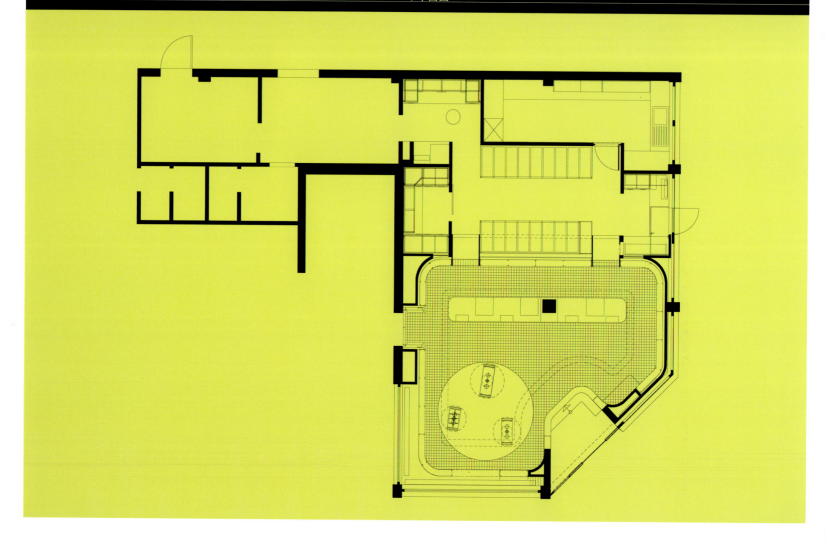

↑ 纯粹的白色墙面 / The pure white walls.
← 空间彩绘细部图 / Space painted detail map.
↓ 精致的货架和墙面的弧线 / Sophisticated arc shelf and wall.

LINE ELEMENTSLINDEN IN S.OLIVER
s.Oliver形象概念店铺

28

【坐落地点】德国慕尼黑；【面积】1672 m²
【设计】Plajer & Franz Studio〔德〕
【材料】墙纸、镀铬板、紫檀木、地砖、地毯
【摄影】Diephotodesigner.De

每一个品牌都有自己最独特的系统，这就意味着，每一个s.Oliver的产品都有一个相同之处。他们虽然被划分到了不同的领域，如男装、女装、童装，但表现出来仍然是s.Oliver这个品牌的精神。这些现象反应在具体的空间设计中，就表现在各个空间的不同与融合这一概念上。针对不同的客户和市场，每个空间都有其独立的设计手法，而各个空间也都结合了经典的s.Oliver的品牌特征，且和建筑元素相结合，形成一个融合的整体。

s.Oliver休闲女装部中，红白相间的色彩搭配突出了它时尚简约的设计精神。白色，可以显出商品的特性，让它们更加引人注目。而红色则可以让空间更显活泼，突出一种节奏感和韵律感，显示出这一系列自由奔放的一面。天花板上红色和白色的水平线条贯穿在空间中，体现了一种流动的美感。休闲男装部则用红色的线条制作了展柜，既延续了女装部的元素又展示了自己的风采。在精选女装部分，设计风格又显不同。整个空间，显的沉稳内敛，有一种低调的美感。深色的镀铬天花与地板相连接，黑色紫檀木搭配米色的块面让空间的色彩显示出了一种内在的平衡。展示台上空长长的丝状装饰安然垂下，精致华丽。墙壁上，s.Oliver代言人的形象随处可见，加深了人们对品牌的认知程度。

Each brand has its own a most unique system, which means that each product has a common s.Oliver. Although they have been divided into different areas, such as men's, women's and children's clothing, but it is still s.Oliver manifested the spirit of the brand. Of these phenomena reflected in the specific areas of space design, manifested in the various spaces and integration of the different concepts. For different customers and markets, each room has its own independent design techniques, while the various spaces have also combines the classic s.Oliver brand characteristics, and architectural elements and combined to form an integrated whole.

s.Oliver Leisure Women's Department, the red and white color with highlights of its stylish minimalist design spirit. White, and underlines the characteristics of goods, so that they are more noticeable. And red, you can make room for even more lively, highlighting a sense of a sense of rhythm and rhythm, showing that this series of free and unconstrained side. Ceiling on the level of red and white lines run through the space, reflects a flow of beauty. Casual men's clothing department with the red lines are produced showcase, both elements of continuity of the Women's Department has demonstrated its own style. In the selection of women's part of the design style has also clearly different. The entire space, explicit steady introverted, there is a low-key beauty. Dark chrome ceiling and floor are connected, black rosewood with beige block surface so that the color space shows an inner balance. Showcase Enron over the long filaments hanging decoration, exquisite ornate. The wall, s.Oliver spokesperson's image everywhere, to deepen people's awareness of the brand.

↑ 鲜艳时尚的店面 / Bright fashion stores.

↑ 室内空间 / Interior space.
↓ 一层平面图 / 1st floor plan.

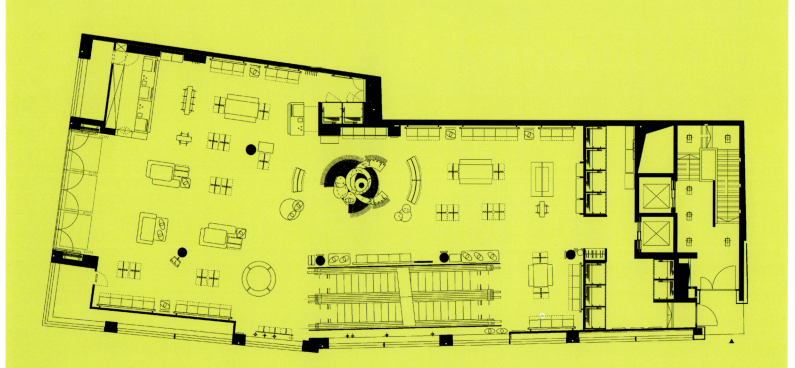

↑ 天花板上红色和白色的贯穿 / The ceiling, red and white throughout.
↓ 二层平面图 / 2nd floor plan.

↑ 精选女装显的沉稳、内敛 / Featured Women's obvious the calm, introverted.
← 男装部的展示架体现了一种动感 / Men's Department reflects a dynamic display.

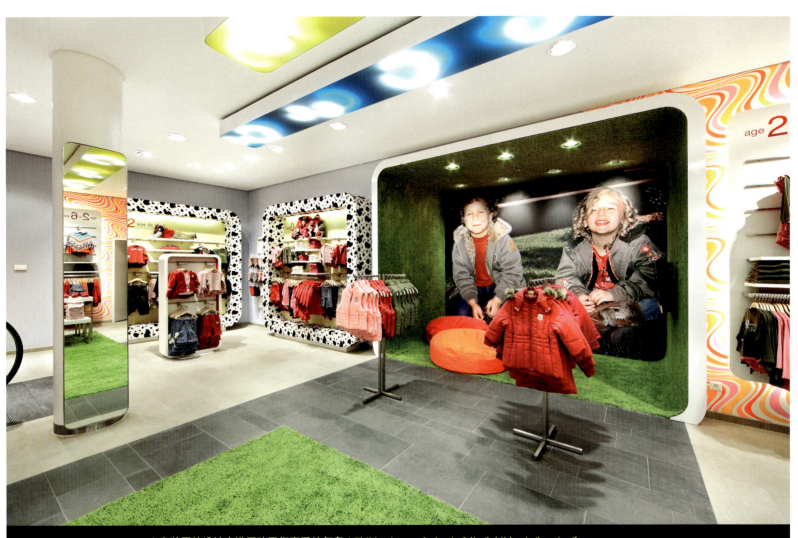

↑ 童装区的设计充满了孩子们喜爱的气息 / Children's area design is full of children's favorite flavor.
← 休闲男装区的设计有一种刚硬豪迈 / Men's leisure area design has a rigid heroic.
↓ 三层平面图 / 3rd floor plan.

ASOBIO SHOP IN SHANGHAI

Asobio店铺设计

29

【坐落地点】上海Channel One购物中心
【面积】1 272 m²
【设计】nendo设计事务所〔日〕
【摄影】Jimmy Cohrssen

设计师为这个空间定义了一个特别的设计主题——"聚焦"。空间在焦点收回刹那间带来深度与连续性的变化，而人们也在光影的对比中体现出了一种梦幻般的购物体验。与此同时，随着空间的纵深感和连续性不断变化加深，店内陈列的商品也在这一刻"不经意"地被聚焦起来。

空间的主调为黑白两色，延续了nendo设计事务所2008年为ASOBIO设计的店铺的色彩风格。而将设计主题定义为"聚焦"，可以说在一定程度上延续了第一家店"图片的工作室"的概念。从第一间店铺的单纯对摄影和摄像器材的模仿，到新的旗舰店的对于镜头变焦的理解，设计师的设计思想也经历了一定意义上的升华。黑色与白色的对比，被灰调的落叶图案所调和，变得更加亲近柔和起来。减弱的对比度反而能够让商品更加突出。

店铺设计使用的材料并不复杂，顶部没有加入任何装饰，干净整洁。这样也可以使空间的设计主题更加突出，同样让顾客们的心思更容易放在店铺的陈列品而非花哨的室内装饰之上。

Designer for this space defines a particular design theme - "Focus." Space in the moment to bring back the focus of the changes in the depth and continuity, while the contrast of light and shadow people are reflected in a fantastic shopping experience. At the same time, with the depth of space and continuity of the ever-changing deepened, store display of goods is also at this moment, "accidentally" been focused up.

The main theme of space for the black and white, and continued the nendo design office in 2008 for the design of the shop ASOBIO color style. The design theme will be defined as "Focus" can be said to a certain extent, continues its first store "photo studio" concept. From the first shop just for photographic and video equipment, imitation, to the new flagship store for the zoom lens of understanding, the designer's design concept has also experienced a certain sense of sublimation. Black and white contrast, was gray leaf pattern of the harmonic tone becomes more gentle closer together. Reduced the contrast and they can make products more prominent.

Shop is designed to use the materials is not complicated, at the top has not joined any decoration, clean and tidy. This would also allow a more prominent space design theme, the same customers on their mind more easily store displays rather than on fancy interior decoration.

↑ 黑色与白色的对比 / Black and white contrast.

↑ 店铺外景 / Shop location.
↓ 简单纯净的质感 / Simple and pure texture.

↑ 地板上单色落叶的图案 / Monochrome leaves patterns on the floor.
↓ 店内的橱窗 / Store windows.

↑ 墙壁的单色落叶图案 / Leaf pattern on the wall color.
↓ 简洁的空间 / Compact space.

↑ 品牌的形象和设计理念完美融合 / Brand's image and design the perfect fusion.
↓ 试衣间 / Fitting room.

SIRIUS SMART SOUNDS PRAGUE
SSSP 唱片连锁店

30

【坐落地点】布拉格，捷克
【面积】78 m²
【设计】Maurice Mentjens
【摄影】Veva van Sloun

布拉格是欧洲炼金术的中心，于是整个设计围绕炼金实验室这一主题展开。以实验室为寓意，主要表达了炼金术科学的一面。从此"炼金术"也成为了Sirius连锁店的基本母题。在其位于Roermond和Eindhoven的分店里，仍然以炼金术这一主题为出发点，分别表达了其精神的一面和炼金的过程。

在Eindhoven分店里，房间中央放置了一块为缤纷色彩所环绕的黑色石头，这一黑色石头在布拉格新店中变成了金色。布拉格新店基本以白色为主，只有柜台例外，由黑色和金色组成。当你进入第一道门的时候，柜台是黑色的，但当你进入下一道门时，柜台便变为金色。这一特殊效果是通过使用特殊的清漆而取得的。虽然空间基本以白色为主，但通过照明的变化却也可以变化出缤纷的色彩。在店中的过道墙上，摆放了很多钟形玻璃容器，其中储藏有药剂是在店中出售的。柜台后面的正方形壁龛不但可以用于放唱片并可以成为店铺的展示墙。

房间尽头的墙上覆盖了玻璃，这让房间显得比较大，而在墙上则设计了一个大DJ台。在DJ台的上方，盘旋着一个大的圆形光环，在圆形光环的中间有一个象征不朽的圆点，两者的结合象征着宏观和微观世界的统一。在空间中，悬空设置的类似横梁的结构，通过在上面安装荧光灯，从而赋予了空间以丰富的色彩变化。

Prague is the center of alchemy in Europe, so the whole design around the themes of alchemical laboratory, laboratory implies, the main expression of the scientific side of alchemy, from "alchemy," Sirius has also become the chain's basic motif, and Eindhoven in their stores located in Roermond, the still the subject of alchemy as a starting point, respectively, expressed his spiritual side and the alchemical process.

In the Eindhoven branch, the one placed in the middle of the room surrounded by colors of black stone, the black stone in the new store in Prague, became a gold. Prague-based stores, mostly in white, only the counter exceptions, by the composition of black and gold. When you enter the first door when the counter is black, but when you enter the next door, the counter will be turned into gold. This special effect is through the use of a special varnish obtained. Although the basic space-white-based, but by changes in lighting, but can also change out the fun colors. The wall of the aisle in the store, placed a lot of bell-shaped glass containers, of which there are pharmaceutical storage in the store for sale. The cabinet behind the counter not only can be used to play a record and can become a store display wall.

Cover the walls of the room at the end of the glass, which make the room seem larger, while in the wall designed a great DJ sets. In the DJ sets at the top, circled with a large round halo ring in the middle of a circular symbol of enduring dot, the combination of macro-and micro-world, a symbol of unity. In space, vacant structures set up a similar beams, through the installation of fluorescent lights in the above, giving the space to enrich the color changes.

↑ Sirius Smart Sounds Prague的入口 / Sirius Smart Sounds Prague entrance.

↑ 房间中的大搁物架中放着各种各样的唱片 / A large room shelf shelving stood in a wide variety of music.
↓ 在DJ台的上方,盘旋着一个大的圆形光环 / In the DJ sets at the top, circled with a large round ring.

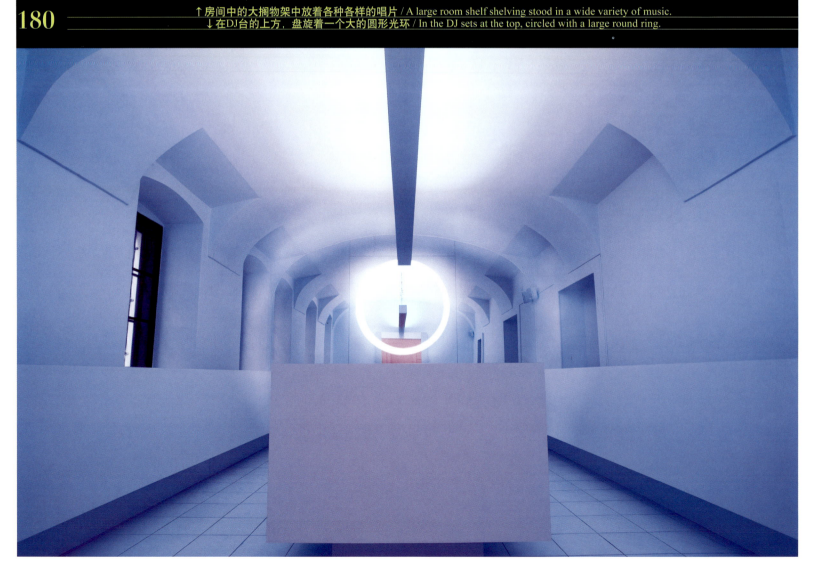

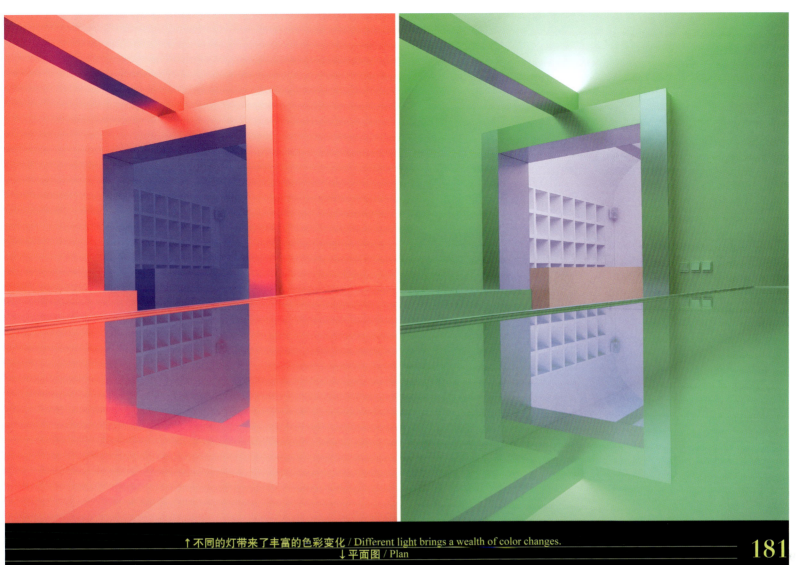

↑ 不同的灯带来了丰富的色彩变化 / Different light brings a wealth of color changes.
↓ 平面图 / Plan

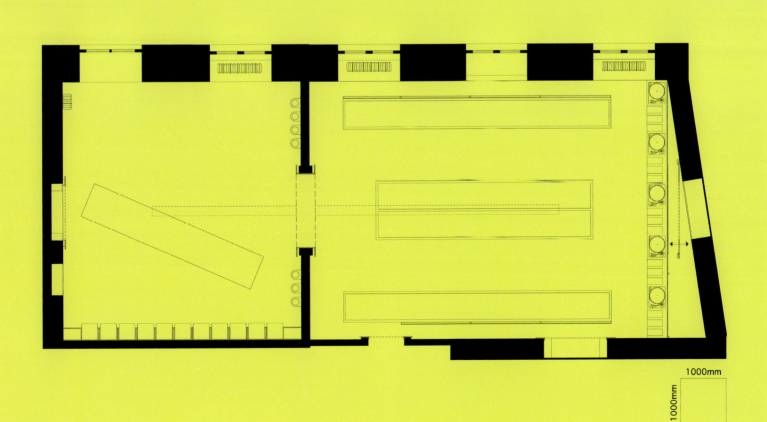

BARBIE
FLAGSHIP STORE
芭比上海旗舰店

31

【坐落地点】上海淮海中路550号
【面积】3 500 m²
【设计】Slade Architecture 〔美〕
【摄影】贾方

旗舰店的外立面宏大美丽，内层是半透明聚碳酸酯材料，外层的平板玻璃印有BIG公司设计的图形。两层玻璃使得整个建筑包裹在透明粉色的糖衣之中。标志着一个属于芭比的瑰丽的梦幻世界。

从一层往上有一部直达三层的电梯，电梯正对建筑的主入口。从三层往上，楼层之间相互联系的是一个旋转楼梯。从功能分布的角度来看，总共六层的空间被分为两大部分，一层至五层是为女孩子们准备的，六层有一个"芭比餐厅"，需要乘坐另外一部电梯才能到达。

三层是顾客乘电梯到达后得到的第一印象。这个空间的布局相对松散宽敞，主要有一个芭比时装的展示区、试衣间。其中，最吸引女孩目光的也许就是旋梯一侧的糖果墙和美容沙龙了。糖果墙上的各色糖果装在一个个定制的玻璃壁柜中，远看就像是墙上的涂鸦一般。在美容沙龙里，银色和粉色的马赛克打造出一个金粉空间，墙上的一面面圆形镜子就像是五彩气泡，里面装的全是美丽梦想。

旋梯而上的四层有一个设计中心，一侧是售卖芭比的零售区。设计师用不同颜色的墙面和地面来区分空间的不同功能。设计中心的外墙同样是玻璃展示区，里面陈列的芭比是不同时期的限量版。紧接着，在五层，设计者为孩子们提供了一片阅读空间，这里陈列着各种有关芭比的各类书籍和画册。

Flagship store's facade magnificent beauty, inner layer is a translucent polycarbonate material, the outer layer of flat glass printed BIG-designed graphics, two layers of glass, making the whole building wrapped in a transparent pink icing into. Marks a magnificent part of Barbie's dream world.

From the floor up there is a direct three-story elevator, elevator is on the building's main entrance. From the three-up, the floor is a link between the rotating staircase. From a functional point of view of distribution, a total of six-story space is divided into two parts, a layer to the fifth floor is for the girls prepared for the six-story there is a "Barbie restaurant."

Three-story elevator to reach the customers obtained after the first impression. The layout of this space is relatively loose and spacious, the main there is a Barbie fashion show area, dressing rooms. Among them, the most attractive girls, eye candy is perhaps the ladder side of the wall and a beauty salon. Candy candy colored wall mounted on a glass cabinet in a customized, seen from afar like a wall of graffiti in general. In the beauty salon, the silver and pink mosaic to create a powder room, a circular mirror on the wall side to side like a colorful bubbles, which are all filled with beautiful dreams.

Four on the ladder but there is a design center, the side of the sale of Barbie's retail district. Designers use different colors to distinguish between the walls and floor space of different functions. Design Center is also the external walls of the glass display area, which displayed a different period of Barbie limited edition. Then, in the five-story, designers provided for children a reading room, where on display are various types of Barbie books and pictures.

↑水晶旋梯 / Crystal ladder.

↑ 大楼夜景 / Building night；↑ 通向三楼零售大厅的手扶电梯 / Leads to the third floor of the retail lobby elevator.
↓ 粉色造型的珍贵芭比娃娃 / The shape of valuable pink Barbie doll.

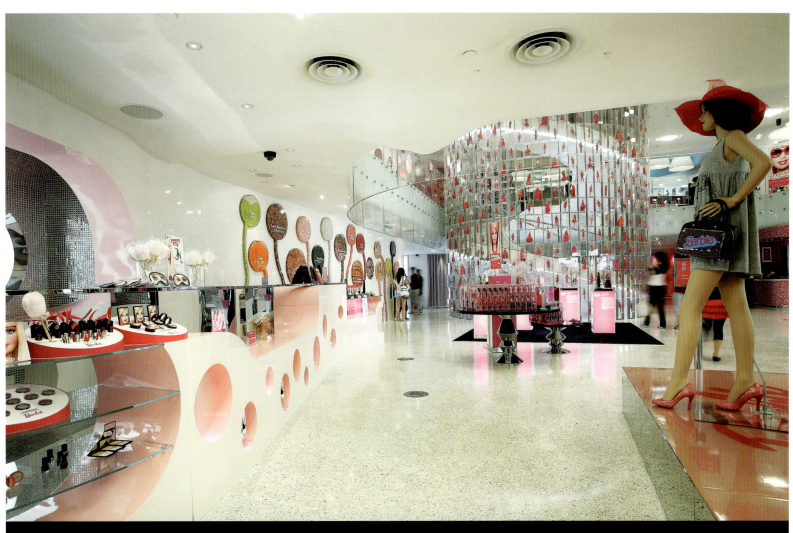

↑ 一楼 / On the first floor.
↓ 一层平面图 / 1st floor plan.

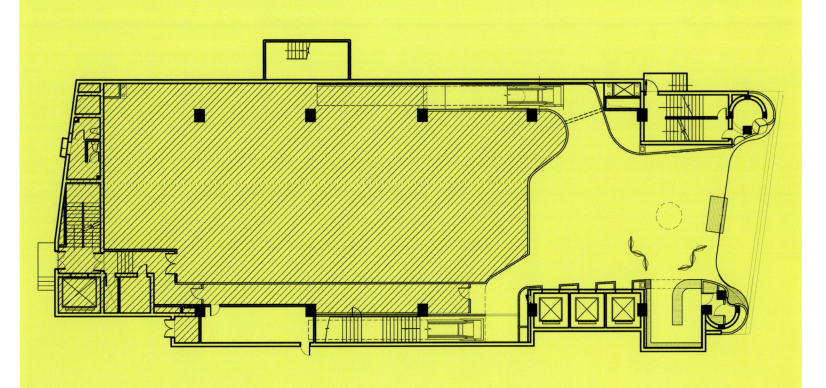

↑ 书籍视听区 / Audio-visual area.
← 时尚服装秀 / Fashion fashion show.
↓ 二层平面图 / 2nd floor plan.

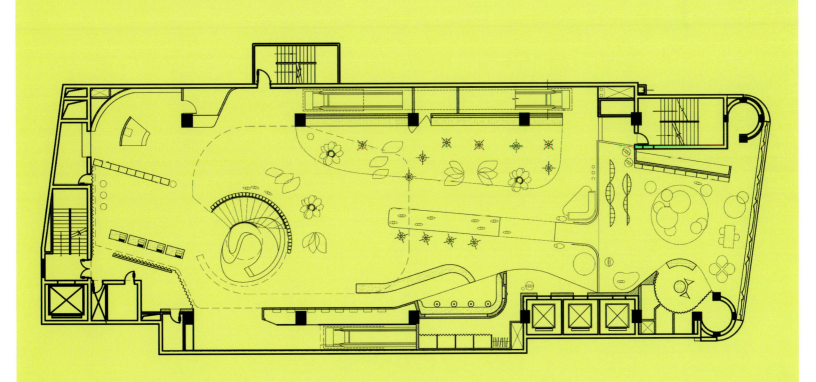

↑ 墙体被充分利用 / The wall has been fully utilized.
↓ 三层平面图 / 3rd floor plan.

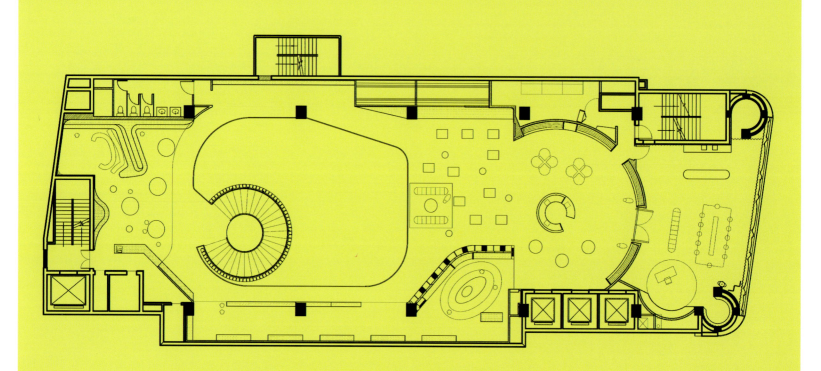

↑ 精选玩具区 / Featured Toy district.
↓ 四层平面图 / 4th floor plan.

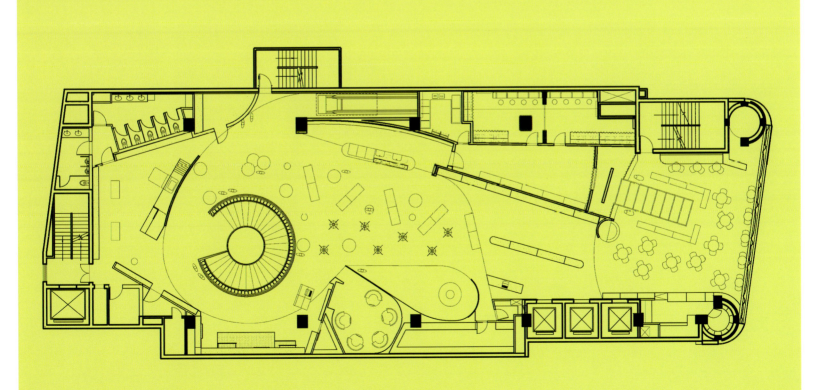

↑ 酒吧区 / Bar area.
← 芭比咖啡区 / Barbie coffee area.
↓ 五层平面图 / 5th floor plan.

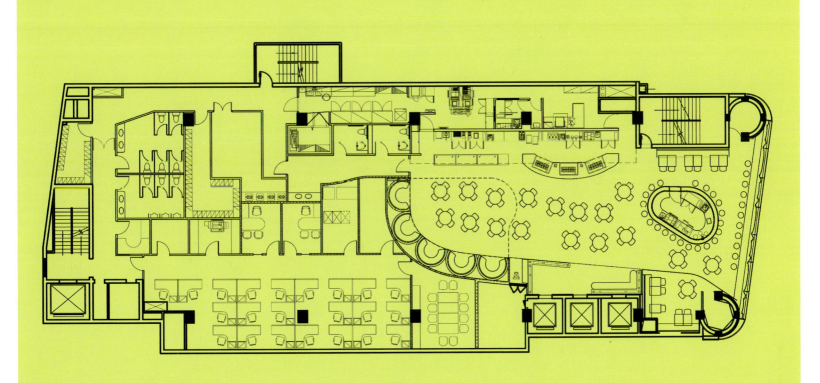

H.N.LIN
DESIGN CENTRE IN SHANGHAI
上海空间美学馆

32

【坐落地点】上海市闵行区七莘路3758号；【室内面积】1楼 2900 m²、2楼 2600 m²、3楼 800 m²
【总体策划、总监】林宪能 Hsien Neng Lin；【参与设计】李玮珉（原有建筑设计）；钟培宁
【主要建材】黑瓦色清水模地、5 cm²实木地板、钢化彩玻
【摄影】OTO Photo

空间美学馆的一层空间被规划为数十家名牌家具的展示区，二层空间被规划为艺文活动区，另有一个位于夹层的现代家具博物馆。

由于原有厂房建筑拥有较高的层高，设计利用原有的空间高度将空间在垂直上作分离，在同一层面创造出从0.45~1.5 m不同高度的展示平台，各个品牌通过高差自然分隔，人们的视线可以穿越整个空间，并清楚地获知自己所在的位置及将要去的方向，创造出通达的视觉体验。

设计在两个位置突破了原有厂房建筑的外壳。在入口处，设计了一个跨7 m的停车入口，从建筑的两个柱子之间延伸，穿过道路直达草坪，由一个工字钢框架和一个品牌标志墙支撑。这个入口的设计引导参观者的视线，既标志着入口，也从现代风格的设计暗示着展馆内以设计为主题的展示风格。另外在空间美学馆的东北侧，原建筑立面被拆除，室内空间延伸到户外，成为一个供未来可使用的咖啡吧区。在二层的博物馆区和艺文活动区，为了满足灵活隔间的使用，与一层的大面积开放式不同，二层空间被分割成两个楼层，引导参观者先进入博物馆区，接着进入不同大小的活动隔间。

Museum's first floor space aesthetic space is planned as dozens of brand-name furniture display area, the second floor space is planned as artistic and cultural activity area, and another one in the mezzanine of the Museum of modern furniture.

Since the original plant has a high-storey buildings designed to use the original height of the space to make space in the vertical separation created at the same level from 0.45 ~ 1.5 m height of the display of different platforms, different brands separated by a natural elevation difference , people's attention to pass through the entire space, clearly informed of their location and to go the direction of creating accessible visual experience.

Designed to break in two locations of the original factory building shell. At the entrance, designed a 7 m of parking across the entrance from the building between the two columns extending across the road directly to the lawn, from an I-beam frame and a wall of support for the brand logo. The design of the entrance of the vision to guide visitors, both marked the entrance, but also from the modern style of design implies a design theme exhibition halls display style. Also in the museum space northeast of aesthetics, the original facade was removed, interior space extending to the outdoors, for future use as a coffee bar area. In the second floor of the museum area and the artistic and cultural activities area, in order to meet the flexible use of compartment, with a layer of a large area of open is different from the second floor space divided into two floors, and guide visitors entered the museum area, and then enter the the activities of different size compartments.

↑ 主入口 / The main entrance.

↑ 灯具展示 / Lighting display.
↓ 挑高与平面的结合 / The combination of high-ceilinged and plane.

↑ 室内空间 / Interior space.
↓ 平面图 / Plan

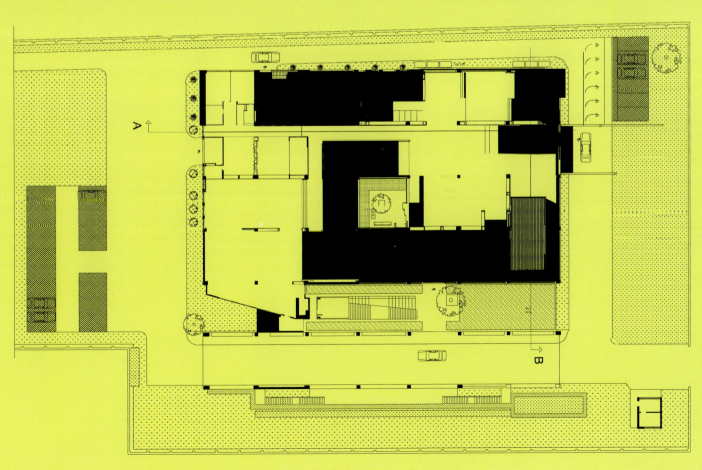

↑陈列台高度的变化，适应于地面不同的标高 / A high degree of change, to adapt to a different elevation on the ground.
↓Cassina文化家具展示区 / Cassina and cultural furniture display area.

↑ 垂直高度凸显空间的气度 / Vertical height highlight the magnanimity of space.
↓ 不同的高度，不同的视线 / Different altitudes, different line of sight.

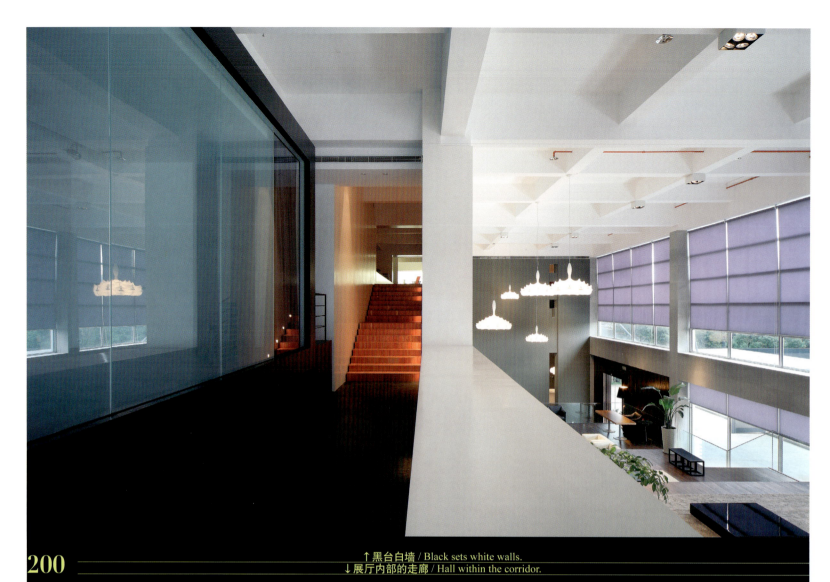
↑ 黑台白墙 / Black sets white walls.
↓ 展厅内部的走廊 / Hall within the corridor.

↑ 典型的博物馆陈列方式 / The museum displays a typical way.
↓ 木地板散发轻松自然之感 / Wood flooring distributed natural feeling relaxed.

Isoleé
FASHION+FOOD+LIFESTYLE
Isoleé时尚生活馆

33

【坐落地点】西班牙马德里Infantas大街19号
【面积】200 m²
【设计】Teresa Sapey建筑事务所
【摄影】Pablo Orcajo

整座馆共分两层：一层包括自助餐厅、熟食超市和商品零售区；休闲室和卫生间位于地下室。建筑的构架通过强调"线"的图形感来展现空间的特性。因此，直线的概念可以视作设计的初始点。

一层，"线"源于主墙面上所画的一棵树的图案。树的分权向各个方向伸展，并在墙面上形成序列状的一道道白色的直线，将各个空间的墙面连接起来。直线之间填充以不同色阶的灰色色带，随着墙面延伸，引导着客人在空间中游走。同样，地下室的卫生间也绘有树的图案，树的分枝成为构建空间特性的最重要的元素。树的造型加强了视觉的紧凑感，同时也成为从此处延伸出去的白色线条的视觉中心。这些"线"绕熟食超市一圈，在零售区又转而变成架子、衣柜或是抽屉。在墙面上呈现的图案形式根据视觉整体性的需要，采用冷与暖、深与浅、曲与直的对比。同样的树的图案造型，在地下室、卫生间内也有呼应，甚至在商店的外立面上，采用黑色滚涂的手法，以体现同样的主题。

The whole museum is divided into two layers: a layer including a cafeteria, cooked food supermarket and retail areas; recreation room and bathroom in the basement. Architecture framework by emphasizing the "line" graphics to show a sense of space features. Therefore, the linear design of this concept was used as the initial point.

First floor, "Line" from the main wall painted a tree pattern. Tree sub-branches of a tree stretching in all directions, and the wall to form a sequence of white-like a straight line, will connect the various spaces of the wall. Fill a straight line between the different Levels of gray ribbon, with the extension of the wall to guide the guests walk in space. Similarly, the basement bathroom is also decorated with tree patterns, tree branches into a spatial characteristics of the most important element. The tree a compact shape to enhance the visual sense, but also extend out from here to become the visual center. These "lines" around the circle cooked food supermarkets in the retail areas in turn becomes a shelf, wardrobe or drawers. The pattern on the wall showing the form of visual integrity according to the needs, using cold and warm, deep and shallow, with curly and straight comparison. The same pattern of tree shape, in the basement, the bathroom also has echoes, even in the store's facade, using the black paint the way to reflect the same theme.

↑入口前行，左边为商品区，右边是吧台 / Front of the entrance line, on the left for the merchandise area on the right is the bar.

↑入口 / Entrance
↓休息区 / Rest area; ↓展示柜 / Display cabinets.

↑服装零售区 / Apparel retail district.
↓服装零售区 / Apparel retail district.

↑↓地下室的休闲区，"树"的形式主题再次出现/ The basement recreation area, "tree" in the form of the theme again.
→地下室的休闲区 / The basement recreation area.

↑↑ 试衣间和洗手池 / Fitting room and hand-washing.
← "树"的主题吧台 / "Tree" theme bar.
↓↓ "男士"和"女士"的洗手间 / "Men" and "Ladies" in the toilet.

KAI YUAN
FOOD COMPANY
开元食品展示中心

34

【坐落地点】中国台湾台北市民权东路；【面积】室内面积：192.72 m²；基地面积：266.97 m²
【设计】甘泰来；【设计公司】齐物设计
【主要建材】玻璃洗石子、白色人造石、乳白色压克力
【摄影】卢震宇

展示中心的正面采用全透明的玻璃，空间里面的内容尽收眼底，这种方法被设计师称为"橱窗概念"。设计师则另辟蹊径，将橱窗放大，通过抽象的几何形体规划出不同的功能空间。

在入口处，左边是几扇具有东方特色的旋转门，穿过旋转门，一个小小的露天庭院展现在眼前，平滑的石阶与一棵来自东南亚的鸡蛋树，共同形成了一个清新的地景。旋转门的上端几经翻转，与主入口上部微斜的雨遮连成一气。从外面看，一个立体的"U"形将内部空间分成三块，入口大门位于中间的部分，设计师有意将地面抬高，下设灯带，自入口进门，一直绵延到空间的深处。

开元食品的LOGO是红白两色，因而设计师也采用了这两种颜色作为空间的主色调。红色的展示墙从入口开始，一直延伸到空间的尽头。展示墙上有若干个大小不一的开口，作展示产品之用。每个开口上都有一个对应的盖子，业主可以根据展示的需要自行变化。设计师在展示墙背后加建了一层薄墙体，上面布满了灯管，由于所展示的产品具有透光性，当开口打开时，白色的强光照射在产品上，展示的效果被进一步强化。另外，展示墙上还隐藏着一道门，通向后面的会议室，由于墙面开口起伏不定的节奏感，它的存在几乎难以被发现。

Exhibition Center's positive all-transparent glass, the contents of all the space inside can be seen, this method is designer known as the "showcase concept." Designers will showcase amplified by the abstract geometry planning a different function space.

At the entrance on the left is a few fans with oriental features revolving door, through the revolving door, a small show in front of an open-air courtyard, a smooth stone steps with an egg tree from Southeast Asia to jointly form a fresh landscape. The upper end of the revolving door several times and turned over to the upper part of the main entrance ramp with the rain cover with these cities. From the outside, a three-dimensional "U"-shaped internal space is divided into three, in the middle part of the entrance door, designers interested in the ground elevation, consists of light zone, since the entrance door, has been stretching into the depths of space.

LOGO food is red and white two colors, so designers have also used these two colors as the main colors of space. Display on the wall there are several different sizes of the openings for displaying product use. Each has a corresponding opening on the lid, the owner can show changes in the needs of their own. Designer behind the display wall, the construction of a thin wall, covered with a lamp, because of the display products are translucent, when the opening is open, the white light irradiation in the products, show the effect of being further strengthened. In addition, the display wall is a hidden door leading to the back of the conference room, wall openings due to fluctuating sense of rhythm, and its existence is almost difficult to be found.

↑ 正面是全透明的玻璃，橱窗的概念引入设计中 / The front is a completely transparent glass, a window introduces the concept of design.

↑ 入口处 / Entrance
→ 入口的接待区 / The entrance reception area.
↓ 入口的艺术墙 / The art of the entrance wall.

↑ 露天庭院 / Open-air courtyard.
↓↓ 地面上镜面不锈钢的积水意象 / Ground water on the mirror image of stainless steel.

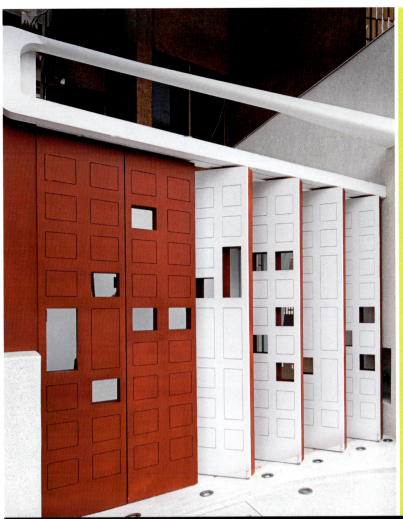
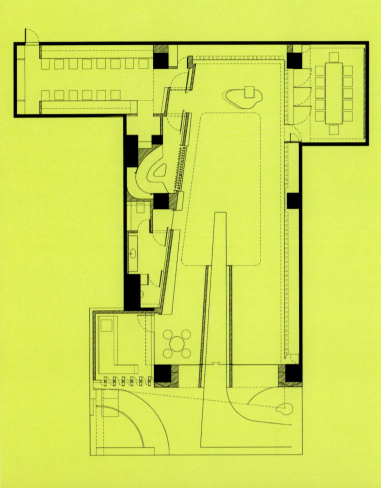

↑平面图 / Plan
↑↓↓红色的旋转门，旋转出不同的景色 / Red revolving door, rotating a different view.

VITRA STORE
Vitra家具零售商店设计

35

【坐落地点】巴西圣保罗；【面积】项目面积：504 m²；建筑面积：250 m²
【设计】Gabriel Kogan〔巴〕；【设计公司】STUDIO MK27
【承包商】pentágono engenharia
【摄影】Nelson Kon Fran Parente

本案中新旧两个平行的建筑经由一个通道相连，设计师创造了一个更大的空间，完成了一个丰富的、不可思议的作品，并将一个全新的VITRA家具的展示厅展现在人们的眼前。清水混凝土一直是当代设计师最钟爱的建筑材料之一，本案也运用了这一低调而质朴的材料。施工方式上，设计师大胆重新启用了曾经一度在民间流行的最原始的建造方式，建筑内外的墙体都没有任何精心粉刷的痕迹。建筑的前立面处开出了一个比例近乎完美的橱窗，低且长，外部的人们一眼就可以看到里面展示的设计作品。

展示空间后部办公室的外立面打破常规，将原本掩藏在墙体之中的钢筋暴露出来，形成了一个如纤维般交织的墙面，极具观赏性，阳光照射下更会出现迷人的光影效果。钢筋粗野的质感和这种精密的交织方式，随意中透着精致的美感，让人眼前一亮。外部的地面用砂粒铺设而成，这些砂粒在建筑的混凝土中也有运用，从某种层面上讲，这也是对建筑本身的呼应。

这个项目的设计包括建筑和庭院两个部分，建筑部分的重要性不言而喻。建造的过程也是一个重生的过程，项目本身的定位是野性主义，让建造者使用一些特殊的表现方式，使建造的过程得以在建筑上真实地反映出来，甚至橱窗和内部的墙体都留下了建造的痕迹，整个建筑以一种狂野的原生态的形式完成。

This case, two parallel old and new buildings connected by a channel, designers created a larger space, completing a rich, incredible work, and a new VITRA furniture showroom display in people's eyes. Fair-faced concrete has been a favorite building material of contemporary designer, one of the case also uses the low-key and rustic materials. Construction methods, the designers boldly re-enabled once the most primitive in a popular way to construct the building inside and outside of the wall do not have any well-painted signs. Building facade at the former out of a near perfect ratio of windows, low and long, people outside can see inside a display design work.

Exhibition space to break the routine back office's facade, originally hidden in the wall among the exposed steel, forming an interwoven, such as fiber-like walls, highly ornamental, sunlight will appear attractive light and shadow effect. Steel rough texture and this mixed precision approach random trace exquisite beauty, it all themselves. Outside the ground with sand flooring, these sand in the construction of concrete also has the use of a certain level of speaking, this is also echoed on the building itself.

The project design includes two parts of buildings and courtyards, building part is very important. Construction process is a process of rebirth, the positioning of the project itself is a wild doctrine to enable the construction to use the performance of some special way, so that the course of construction to be true in architecture reflected, and even windows and interior walls were left traces of the construction of the entire building in a wild form of the completion of the original ecology.

↑ 夜色下的纤维般交织的墙面 / Night under the walls of fiber-like interwoven.

↑ 建筑的前立面处的完美的橱窗 / Before the building facade at the perfect showcase.
↓ 朦胧的影子 / Hazy shadow.

↑ 它区别于其他常见的清水混凝土建筑 / It is different from other common concrete construction.
↓ 平面图 / Plan

↑ 设计组合 / Design combination.
← 内墙上保留了工人粉刷的痕迹 / Retention of workers within the walls painted the scene.
↓ 阳光照射下更会出现迷人的光影效果 / Sunlight will appear attractive lighting effects.

HONDA
EXHIBITION HALL
广州本田企业形象展示厅

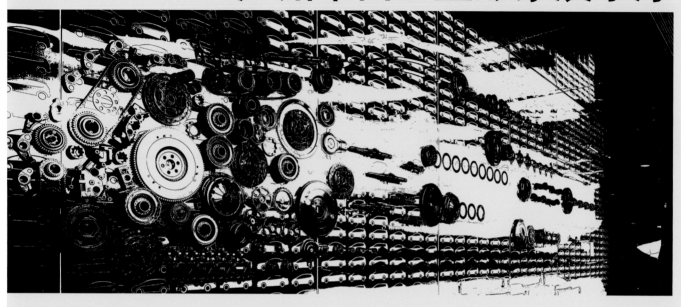

36

【坐落地点】广州天河区天河路；【面积】约1 900 m²
【设计】盛彦明
【设计公司】广州市雅哲工程设计有限公司
【摄影】贾方

整个空间划分为接待前厅、车型展示厅、洽谈区、体验区、企业文化及切割车展示厅等。设计师为空间设计了两条参观主轴线，轴线的终端均为设计亮点，吸引受众走向展示厅的纵深区域，很好地解决了展场原来整体性不够的问题。

一进入展厅，视野的尽头就是一款炫目的概念车型。黑色起伏的镜面天花从大门位置一直延伸到这款车的背景上，连绵不绝的黑色镜面天花和墙身凹凸宛如黑色钻石的造型，加上蓝色LED的点缀，营造出超现实的氛围，同时亦给人带来一种极富理性的感觉。要配合镜面将好的空间效果表达出来，灯光的设计也很重要。不同区域的不同灯光设计，配合空间的层次变化，让整个设计传达的效果更好。在概念车前，有一幅由广本车款内部零件组合成的墙面装饰，大大小小的零件，按照真车配件位置摆放，在增加了空间趣味性、互动性的同时，也显示了该汽车品牌的专业性。

休息区简洁大方的白色沙发，让顾客可以在轻松的氛围里选购心仪车款。再走进展厅内部，一块块洁白宽大的投影白幕，投幕影片详细地为顾客介绍该汽车品牌的文化内涵。极简的空间设计配合镜面效果，让这个文化展示区像一个小型影剧院。

The whole space is divided into reception hall, car showroom, negotiation areas, Experience, corporate culture and cutting car showrooms, etc.. Designer for the space design of the two to visit the main axis, the axis of the terminal are designed to highlight and attract audience into the depth of showroom area, a good solution to the original integrity of the exhibition is not enough problems.

Entered the exhibition hall is a dazzling vision of the end of the concept car, a black undulating ceiling mirror position from the main entrance has been extended to the car, the background of endless black ceilings and walls of the concave-convex mirror is like a black diamond modeling, coupled with a decorative blue LED to create ultra-realistic atmosphere, but also brings a very rational feeling. To meet the mirror effect will be expressed in a good space, lighting design is also important. In different regions of different lighting designs with space-level changes, so that the entire design to convey better. In the concept car before, there is a paragraph from the wide internal parts of the car combined into a wall decoration, large and small parts, car parts in accordance with true display of increasing the number of space fun, interactive, while also shows The car brand professional.

Lounge simple elegant white sofa, so that customers can buy in a relaxed atmosphere crush cars. Re-entered the exhibition hall within a block of white large white projection screen, video screen cast details for the customer to introduce the car brand's cultural connotations. Minimalist interior design with the mirror effect, so that this culture is like a mini-theater display area.

↑ 概念车 / Concept car.

↑ 接待台动感的设计 / Front Desk dynamic design.
↓ 休息区 / Rest area.

↑镜面让汽车更加炫目 / Mirror make cars even more dazzling.
↓镜面让汽车更富时尚感 / Mirror make cars more stylish.

↑ 文化展示区 / Cultural display area.
← 大型的木质装置 / Large-scale wood installations.
↓ 平面图 / Plan

DESIGN JOYS
IN MORALE FURNITURE
北京木皇家具

37

【坐落地点】北京搜狐尚都
【面积】918 m²
【设计】yeonhee Keum〔韩〕
【摄影】贾方

空间整体以白色为底,但却不甘于平淡。弧线的元素被大量运用,使白色一下子生动起来,并起到一定空间引导的作用。而采用粉红色、红色、橙色等颜色组成的公司标志图案镶嵌其中,在大面积的冷色中点入浓烈的暖色,画龙点睛。在形态方面,一些非日常性的元素。加上不同的材料和灯光的巧妙运用,使图案的效果最大化,给人带来意想不到的惊喜。

当影响空间的大部分要素得到简化,惟一剩下的要素的特质会更加鲜明。白色人造石和白色花样地坪漆的使用使得接待处的设计感得到了最大地体现。而接待处色彩鲜艳的地毯,恰到好处地缓解了空间里可能会产生的冷冷的感觉。接待处的桌子延续地面地毯的颜色、图案和材质,和地毯连成一体;再加上接待处桌子与柱子的结合,家具、墙壁与地板连接成一体,整个接待空间形成了强有力的起伏感。

在展示区,自然元素的母题得到了更大限度的发挥。座椅展示区的墙面有一面巨大的镜子,设计师把镜子上可以透光的部分处理成自然的树形和蝴蝶图案。白天产生自然光的效果,夜晚则运用上下部的间接灯,使图案效果最大化。为了使自然光不过多地透过办公区照射到展示区,设计师使用以"乐趣"为主题的玻璃墙。

At the end of a white space as a whole, but not dull. Arc elements have been widely used to make white suddenly to life, and to play a leading role in space. The use of pink, red, orange and other color mosaic pattern consisting of company logo which, in the large area of the mid-point into the thick warm cool and focus. In patterns, some non-routine nature of the elements. Coupled with different materials and the clever use of lighting, so that pattern to maximize the effect, brings unexpected surprises.

When the impact of space most of the factors have been simplified, the only remaining elements of nature will be more vivid. White artificial stone and white pattern makes the use of floor paint design of the reception area has been the greatest sense of reflection. The reception colorful carpet, just right to ease the space may produce a cold feeling. Continuation of the reception desk on the ground carpet colors, patterns and materials, and carpets continuum; together with the reception desk and the combination of columns, furniture, walls and floors are linked into one.

In the display area has been the theme of natural elements play a greater extent. Seat display area of the wall there is a giant mirror, the designer can put the mirror into a translucent part of the deal with the natural tree and butterfly patterns. Produce the effect of natural light during the day, at night the use of upper and lower parts of the indirect light, so that patterns to maximize effectiveness. However, in order to allow natural light through the office area and more exposure to the display area, designers use of "fun" as the theme of the glass wall.

↑ 顾客等候区 / Customer waiting area.

↑ 从入口看接待台 / From the entrance to see Front Desk
↓ 接待台动感的设计 / Front Desk dynamic design.

↑ 接待处后面是与客户的交谈区 / Reception followed by conversation with the customer area.
↓ 平面图 / Plan

233

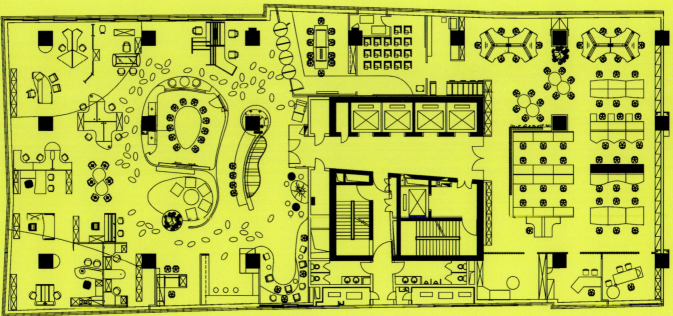

↑ 展示区地面被设计为起伏的两层 / Display area on the ground is designed to be ups and downs of the two.
← 墙面是Hermanmiller的Image wall / The wall is Hermanmiller the Image wall.
↓ 像咖啡厅的空间 / Like the coffee shop space.

↑用绿色和蓝色系设计的橱柜 / Green and blue are designed cabinets.
↓墙面展示Fabric、Laminate、steel等材料 / Walls of the cafe Fabric, Laminate, steel and other materials.

↑培训区 / Training area.
↓会议室 / Conference room.

LCX
DEPARTMENT STORE
Lcx百货公司

38

【坐落地点】香港铜锣湾
【面积】5 000 m²
【设计】莫咏诗
【主要用材】黑钢、亮面不锈钢

大厦的入口处及楼板都偏低，全层柱子及承重墙较多。所以这次设计的首要任务是重新给予顾客一个独特的入口，一个全新视觉的购物空间，令顾客不再局限于原有的建筑规限内。

设计重点在大门入口处，首先楼板经过了改造，把先前只有一层高的入口退后3 m，楼板及横梁被拆掉，重新加固以全高落地玻璃出现，配以闪动的LED电子色灯及亮面切割的不锈钢板块。统一的尺寸，令自然光线互相折射，加上LED发出的独特视觉效果，同时反射在这大门的入口处，营造了非常耀眼触目的视觉感受。另一条LED色带也围着整座建筑物，在设定好的时间内，变出多种颜色效果，使整个店的内外也明亮起来。

主入口天花处设计了多部等离子银幕，配以极具时代感的图案及音乐，不停转动及变幻出多种视觉效果，大型投射幕墙展示Cat Walk Show。为配合LCX Fashion Walk集时尚与艺术于一身的形象，"X"也是这项目的设计元素，除了店内以TrendX、aXessories、CaXual Chic、BioXoms和Cool Xene Cuisine等5个购物区组成外，二层入口处以不同大小的"X"交叠并投射在中间的艺术交流区内，表达出LCX对时装与艺术的独特意念，成为真正以Fashion X Art为题的百货公司。全场灯光系统经过特别处理，特别订制不同长度的光管支架组合，所有灯光色温经过严格限制，务求视觉效果一致。

The building entrance and the floor are low, full-floor columns and load-bearing walls more. So this first task is to re-design to give customers a unique entrance, a new vision of shopping space, so that customers no longer confined within the original architectural restrictions.

Designed to focus on the main entrance, first floor after a transformation, the previously found only in the entrance of a layer of high-back 3 m, floors and beams have been removed, re-strengthening in full high-floor glass appears accompanied by flashing LED E-Color Light and the bright surface of stainless steel plate cutting. Uniform size, so that natural light refraction of each other, coupled with the unique visual LED issue, while reflection in the door at the entrance, creating a very bright eye-catching visual experience. The other is also a LED ribbon around the whole building, in the predetermined period of time, change out the effect of a variety of colors so that the whole shop light up outside it.

Smallpox at the main entrance has designed more than plasma screens, accompanied by a very contemporary design and music, ever-changing rotation, and a variety of visual effects, large-scale projection wall display Cat Walk Show. To tie in with LCX Fashion Walk set of fashion and art in one's image, "X" is also the design elements of this project, the two-story entrance impose different sizes of "X" in the middle of overlap, and artistic exchanges between projected area, expressed LCX pairs of fashion and art. Overall lighting system is special treatment, particularly in light of different lengths custom tube combinations, all the lighting color temperature through the strictly limited to ensure that the same visual effect.

↑入口处 / Entrance

↑ 闪烁的LOGO / Flashing LOGO.
↓ 一层平面图 / 1st floor plan.

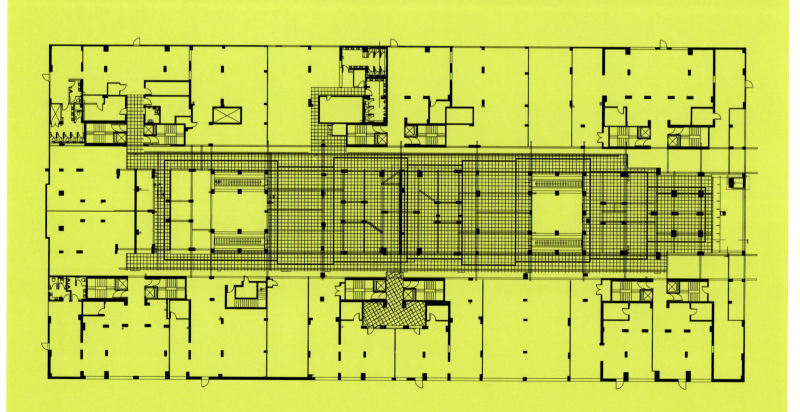

↑室内空间 / Interior space
↓二层平面图 / 2nd floor plan.

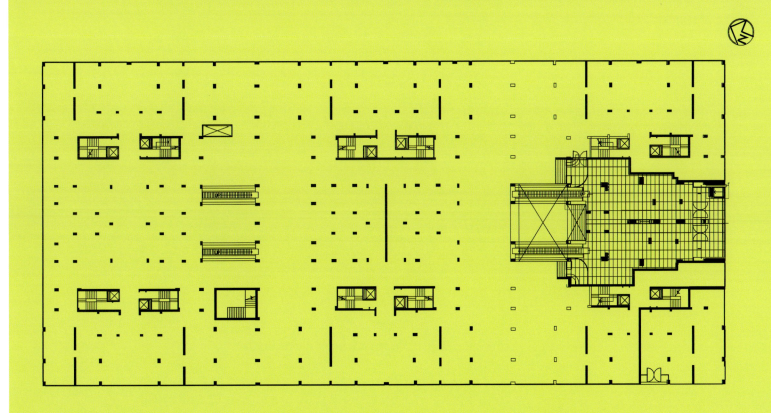

↑ 二层入口处以不同大小的"X"交叠 / Two-story entrance with different sizes of "X" overlap.
← 内部的风格以简洁为主 / Mainly within the style concise.
↓ "X"是整个项目的设计元素 / "X" is the entire project design elements.

C_42 CITROEN FLAGSHIP SHOWROOM
雪铁龙C_42巴黎旗舰店

【工程名称】 雪铁龙C_42巴黎旗舰店
【坐落地点】 法国巴黎 Les Champs Elysées
【面积】 1 200 m²
【摄影】 Philippe Ruault

旗舰店临街的立面运用了三角形、菱形和V字形的玻璃三维拼接，加之色彩上的变化，使一个立体的雪铁龙双箭头标志跃然而出，不经意间强化了品牌的形象。而最让人称奇的是，临街的玻璃墙和建筑顶面、后面的墙体连成一体，是一片扭曲的玻璃体。设计师的这个想法来源于雪铁龙的轮胎，那带有凹槽的橡胶外胎展开后可以任意弯曲，而上面的图案正是雪铁龙的标志——双箭头。设计师通过多个建筑模型，才最后确定了这个方案，而这种扭曲的玻璃造型也给后面的施工带来了很大的难度。

室内当然是以展示汽车为主，设计师把如何最好地体现展示功能作为设计要达到的最重要的目的。既要最合理地利用空间展示，又要让参观者最直观地欣赏展品，设计师最终确定了用螺旋线的方式盘绕而上，作为展台。

每层都摆放了一辆雪铁龙的经典车型，而平台上方的玻璃锥体模仿了钻石的切割方法，分别从100多个方向反射灯光。玻璃锥体下的汽车被照耀得光彩夺目。参观者可以沿着盘旋的楼梯向上一层一层参观。Gautrand希望她的这件作品并不仅仅是一个简单的汽车旗舰店，而像是一个博物馆，一个文化建筑，人们能在这里流连忘返，静静欣赏每一件作品。

Flagship store to use the street facade of the triangle, diamond, and V-shaped three-dimensional mosaic of glass, combined with the change in color, so that a three-dimensional double-arrow logo Citroen skyrocketing inadvertently reinforce the brand's image. The most surprised at is that the street top surface of the glass walls and buildings, behind walls fused, is a distortion of the vitreous. The idea comes from the designer's Citroen tires, then with a groove after the commencement of the rubber tire can be an arbitrary curved, while the above pattern is the Citroen logo - double arrow. Designers through multiple building models, before finalizing this program, which distort the shape of the glass behind the construction also brings a great deal of difficulty.

Of course, show car interior is the main designer of how best to reflect the display capabilities as a design to achieve the most important purpose. It is necessary to display the most rational use of space, but also allow visitors to enjoy the exhibits of the most intuitive, designers to finalize the way of using spiral coil and, as a booth.

Each layer is placed in a classic Citroen cars, while the platform at the top of the glass cone imitation diamond cutting methods, respectively, the direction of reflection from more than 100 lights. Glass cone under the car was too dazzling shine. Visitors can follow the spiral staircase up level by level to visit. Gautrand hope that her work and this is not just a simple car flagship store, rather like a museum, a cultural construction, it is to be here will want to return quietly to enjoy each piece.

↑ 旗舰店外观 / Flagship store appearance.

↑↑ 设计师的灵感来源于橡胶的汽车外胎 / The designer's inspiration comes from car tire rubber.
← 入口接待 / Entrance reception.
↓ 展台示意图 / Showcase diagram.

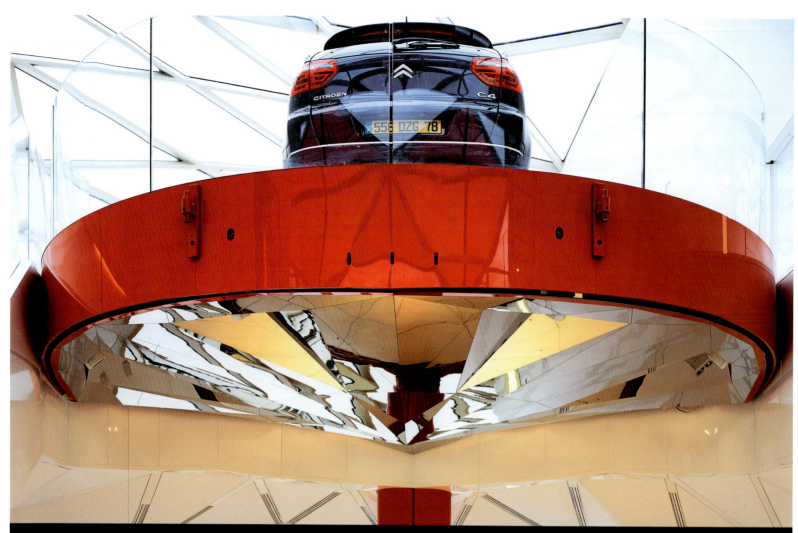

↑圆盘展台直径6 m / Showcase of the diameter of 6 m.
→展台自上而下共8层 / Showcase a total of eight layers from top to bottom.
↓展台顶部的锥形玻璃和室外的菱形玻璃相互呼应 / Stand at the top of the cone-shaped diamond-shaped glass, glass and outdoor each other well.

↑ 正立面玻璃细部 / Glass facade detail; ↑ 顶面玻璃细部 / Top surface of the glass detail.
← 内部的风格以简洁为主 / Mainly within the style concise.
↓ 扭曲的玻璃衬托出老爷车的不凡身价 / Distortion of the glass to set off an expensive car.

HOME DELUXE
TAIPEI SHOWROOM
弘第厨具台北展馆

40

【坐落地点】中国台湾台北市长春路447号1层2层；【面积】一层为285 m²，二层为524 m²
【设计】胡硕峰；【设计单位】胡硕峰建筑工作室 大墙演绎设计有限公司
【主要用材】铝格栅、铁板、水泥板、亮面烤漆、原木松、发色钢板
【摄影】邹昌明

本案的设计哲学是，将现代的简约风格应用在一种极具现代性且有高度质感的空间中，并且在许多住宅、建筑、展示空间、办公室中也是以这样的概念进行设计的。另外，再加上极细致且考虑周详的细部设计，任何空间均能展现一种雍容大气的现代风格。

整个展馆以低限冷冽的黑灰白色调为主题，并使用金属和原木两种冷暖对比的主材料，在一楼刻意营造出一种空旷感以陈列德国顶级Bulthaup厨具。二层为德国LEICHT厨具展场及企业的办公总部，设计师用不同的天花板材料与细部区分出不同的空间机能，创造出一种既分割又连绵不断的开放式空间。

主色调以黑、白、灰为主为本设计带来了一种强烈的几何与现代风格，而位于展示区上方的大木桁架结构则为本案植入了一种现代性中的地域或村庄的风味。这些象征空间桁架的大木结构，一来带来对天花吊顶空间的趣味性和穿透性（侧面有既有的高窗，质感并不好，刚好用木桁架遮住了）；另一方面，把一种旧仓库或厂房的时空元素移置到这个展场，构造出一种与精致产品的对比与趣味。两者共同构造了一个为顶级如手工艺般的厨具品牌的展示舞台。

The case design philosophy is to use a modern minimalist style in a very modern and have a high degree of texture space, and in many homes, buildings, exhibition space, the office is also based on the concept of such a design. In addition, coupled with a very detailed and well thought out detail design, any space can show a kind of grace and modern style of the atmosphere.

The whole complex, the theme of black-gray tone, and use the well-being of two kinds of metal and wood contrast to the main material on the first floor deliberately create a sense of open space in order to display the top German Bulthaup kitchen. The second floor exhibition in Germany LEICHT kitchenware and corporate office headquarters, designers with different ceiling materials and detailing to distinguish different spatial function, create a partition then both continuous open space.

Main colors black, white, gray-based based design has brought a strong geometric and modern style, while the display area is located in the top of the wooden truss structure was implanted in the case of a modernity in the region or village flavor. These symbolic space truss of the large wooden structure, a ceiling space to bring fun and penetrating; the other hand, to an old warehouse or factory space-time elements have been relocated to this exhibition is constructed out of a kind and refined Comparison and interesting products. Both co-constructed as a top kitchen brands, such as arts and crafts like the exhibition stage.

↑ 厨具展示区 / Kitchen display area.

↑↑展厅上方的大木桁架结构 / Hall at the top of the big wooden truss structure.
↓线条感突出的设计 / Sense of prominent design lines.

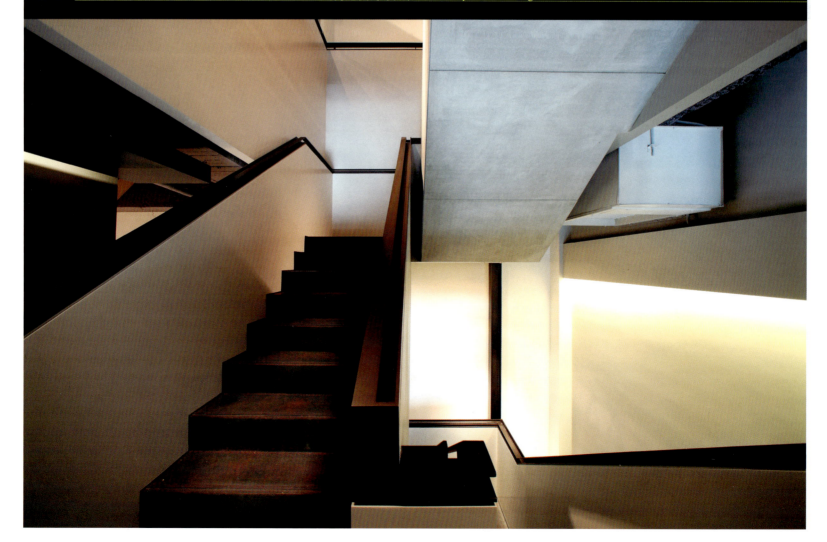

↑ 简洁的黑白灰色调 / Simple black and white gray tone；↑ 回转而上的楼梯 / Rotary and on the stairs.
↓ 平面图 / Plan

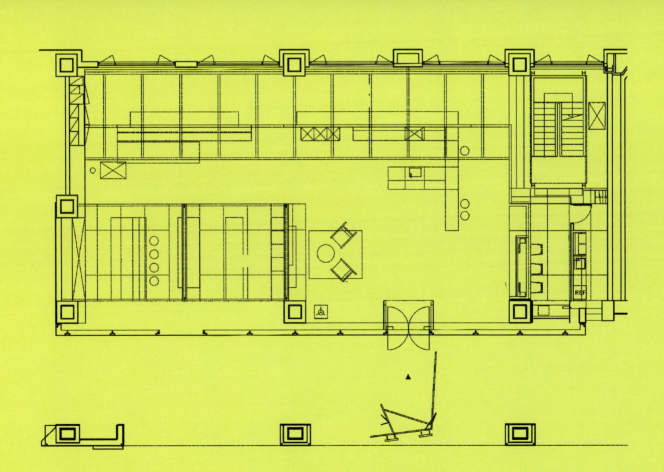

特别推荐 HIGHLY RECOMMENDED

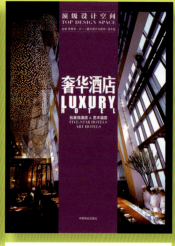

顶级设计空间 I——情调餐厅
ISBN 978-7-5038-5798-0
印装：四色精装
定价：248.00

顶级设计空间 II——纯粹商店
ISBN 978-7-5038-5797-3
印装：四色精装
定价：248.00

顶级设计空间 III——创意办公
ISBN 978-7-5038-5803-1
印装：四色精装
定价：248.00

顶级设计空间 IV——奢华酒店
ISBN 978-7-5038-5802-4
印装：四色精装
定价：248.00

住宅字典
住宅立面造型分类图集. 1
ISBN 978-7-5038-5756-0

住宅字典
住宅立面造型分类图集. 2
ISBN 978-7-5038-5755-2

住宅字典
住宅立面造型分类图集. 3
ISBN 978-7-5038-5754-6

香港日瀚国际文化传播有限公司编
出版：中国林业出版社
印装：四色平装
开本：218mm×336mm
版次：2010年1月第1版
印次：2010年1月第1次
单册印张：20.25
单册定价：198.00

室内设计新作（上下卷）
张青萍 主编
ISBN 978-7-5038-5435-4
开本：230mm×300mm
页码：700
印装：软精装
定价：558.00元
出版时间：2009年6月

顶级样板房 I
张青萍 孔新民 主编
ISBN 978-7-5038-5709-6
开本：230mm×300mm
页码：360
印装：精装
定价：288.00元
出版时间：2009年10月

联系单位：中国林业出版社
地址：北京西城区德内大街刘海胡同7号
邮编：100009
销售客服：13641384559
出版客服：13810400238

网络支持:www.onetopspace.com